HUGH MILLER,
THE OLD RED SANDSTONE:
THE TEXT

Edited by
MICHAEL A. TAYLOR
AND
RALPH O'CONNOR

Published in 2023 by NMS Enterprises Limited – Publishing
a division of NMS Enterprises Limited
National Museums Scotland
Chambers Street, Edinburgh EH1 1JF

National Museums Scotland would like to thank
the University of Aberdeen and The Friends of Hugh Miller
for their generous donations towards this publication.

British Library Cataloguing in Publication Data
A catalogue record of this book
is available from the British Library.

Cover design: Mark Blackadder
Front cover images: the River Findhorn near Sluie. Photo: Ralph O'Connor;
Osteolepis macrolepidotus, NMS G.1859.33.1218. Image © National Museums Scotland.
Back cover image: detail from an engraving by Reverend Drummond of Edinburgh
after photograph by James Good Tunny, *c* 1855. Image © National Museums Scotland.
Internal design by NMS Enterprises Limited – Publishing

Printed and bound in Great Britain by Bell & Bain Ltd.
This product is made of material from well-managed forests and other controlled sources.

For a full listing of related NMS titles please visit:
www.nms.ac.uk/books

The Old Red Sandstone

or New Walks in an Old Field

Hugh Miller

EDITED BY

Michael A. Taylor

and

Ralph O'Connor

CONTENTS

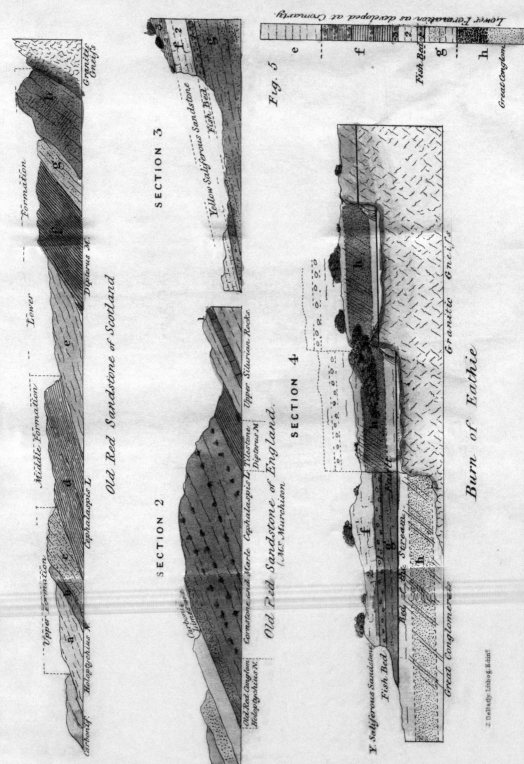

SECTION 1

Upper Formation — Middle Formation — Lower — Formation

Carbonif.ª   Holoptychius N.   Cephalaspis L.   Dipterus M.   Granitic Gneiss

Old Red Sandstone of Scotland

SECTION 2

Old Red Sandstone of England.
(Mr. Murchison.)

Old Red Conglom.   Cornstone and Marle   Cephalaspis L.   Tilestone   Upper Silurian Rocks.
Holoptychius N.   Carbonif.ª Limestone   Dipterus M.

SECTION 3

Yellow Saliferous sandstone — Fish Bed

Fig. 5

Lower Formation as developed at Cromarty

e —
f —
Fish Bed
g —
h —
Great Conglom.

SECTION 4

Y. Saliferous Sandstone.   Fish Bed   Bed of the Stream   Granitic Gneiss

Great Conglomerate

Burn of Eathie

J. Gellaty, Lithog. Edinª

Frontispiece: originally printed as a hand-coloured fold-out plate, with the fold-out opening towards the left, and the long axis horizontal with respect to the book itself. A colour reproduction is provided at Figure 2. Miller explains the individual sections on p. xxi, with the explanation of his fig. 5 coming at the end of the explanation of Section 1.

OLD RED SANDSTONE;

OR

NEW WALKS IN AN OLD FIELD.

BY

HUGH MILLER.

EDINBURGH:

JOHN JOHNSTONE, HUNTER SQUARE.

MDCCCXLI.

JOHNSTONE AND FAIRLY, PRINTERS, EDINBURGH.

RODERICK IMPEY MURCHISON, Esq. F.R.S.

&c. &c. &c.

PRESIDENT OF THE GEOLOGICAL SOCIETY.

In the autumn of last year I sat down to write a few geological sketches for a newspaper ; the accumulated facts of twenty years crowded upon me as I wrote, and the few sketches have expanded into a volume. Permit me, honoured Sir, to dedicate this volume to you. Its imperfections are doubtless many, for it has been produced under many disadvantages ; but it is not the men best qualified to decide regarding it whose criticisms I fear most ; and I am especially desirous to bring it under your notice, as of all geologists the most thoroughly acquainted with those ancient formations which it professes partially to describe. I am, besides, desirous it should be known, and this, I trust, from other motives than those of vanity, that when prosecuting my humble researches in obscurity and solitude, the present President of the Geological Society did not

---

Murchison: see p. 19; **Geological Society**: i.e. of London, founded 1807.

deem it beneath him to evince an interest in the results
to which they led, and to encourage and assist the inquirer
with his advice.   Accept, honoured Sir, my sincere thanks
for your kindness.

Smith, the father of English Geology, loved to remark
that he had been born upon the Oolite,—the formation
whose various deposits he was the first to distinguish and
describe, and from which, as from the meridian line of the
geographer, the geological scale has been graduated on
both sides.   I have thought of the circumstance when, on
visiting in my native district the birth-place of the author
of the *Silurian System*, I found it situated among the more
ancient fossiliferous rocks of the north of Scotland,—the
Lower Formation of the Old Red Sandstone spreading out
beneath and around it, and the first-formed deposit of the
system, the Great Conglomerate, rising high on the neighbouring hills.   It is unquestionably no slight advantage to
be placed, at that early stage of life when the mind col-
lects its facts with greatest avidity, and the curiosity is
most active, in localities where there is much to attract
observation that has escaped the notice of others.   Like
the gentleman whom I have now the honour of addressing, I too was born on the Old Red Sandstone, and first
broke ground as an inquirer into geological fact in a formation scarce at all known to the geologist, and in which
there still remains much for future discoverers to examine

Smith: William Smith (1769–1839), civil engineer and geologist, had only just died, belatedly honoured by the Geological Society of London (Torrens 2001, 2003, 2007). He was born on the 'Oolite' in Oxfordshire and worked out the sequence of geological strata in much of England, using both characteristic fossils and field observations of the relative positions of strata, starting with the 'Oolite' and working up and down the sequence from there. He drew up his conclusions in a series of maps of southern Britain; **author**: Murchison, born at Tarradale House at the south end of the Black Isle. Miller worked nearby as a stonemason in 1821.

and describe. Hence an acquaintance, I am afraid all too slight, with phenomena which, if intrinsically of interest, may be found to have also the interest of novelty to recommend them, and with organisms which, though among the most ancient of things in their relation to the world's history, will be pronounced new by the geological reader in their relation to human knowledge. Hence, too, my present opportunity of subscribing myself, as the writer of a volume on the Old Red Sandstone,

Honoured Sir,

With sincere gratitude and respect,

Your obedient humble Servant,

HUGH MILLER.

Edinburgh, May 1, 1841.

---

the most ancient of things: the oldest fossils then known.

PREFACE.

NEARLY one-third the present volume appeared a few months ago in the form of a series of sketches in the *Witness* newspaper. A portion of the first chapter was submitted to the public a year or two earlier, in *Chambers' Edinburgh Journal*. The rest, amounting to about two-thirds of the whole, appears for the first time.

Every such work has its defects. The faults of the present volume,—faults all too obvious, I am afraid,—would have been probably fewer had the writer enjoyed greater leisure. Some of them, however, seem scarce separable from the nature of the subject : there are others for which, from their opposite character, I shall have to apologize in turn to opposite classes of readers. My facts would, in most instances, have lain closer had I written for geologists

submitted: Miller 1838a, in part.

exclusively, and there would have been less reference to familiar phenomena. And had I written for only general readers, my descriptions of hitherto undescribed organisms, and the deposits of little-known localities, would have occupied fewer pages, and would have been thrown off with perhaps less regard to minute detail than to pictorial effect. May I crave, while addressing myself, now to the one class and now to the other, the alternate forbearance of each?

Such is the state of progression in geological science, that the geologist who stands still for but a very little must be content to find himself left behind. Nay, so rapid is the progress, that scarce a geological work passes through the press in which some of the statements of the earlier pages have not to be modified, restricted, or extended in the concluding ones. The present volume shares, in this respect, in what seems the common lot. In describing the *Coccosteus* (page 53), the reader will find it stated that the creature, unlike its cotemporary the *Pterichthys*, was unfurnished with arms. Ere arriving at such a conclusion I had carefully examined at least a hundred different *Coc-costei* ; but the positive evidence of one speci-men outweighs the negative evidence of a hun-dred ; and I have just learned from a friend in the north (Mr Patrick Duff of Elgin), that a *Coccosteus*, lately found at Lethen-bar, and now in the possession of Lady Gordon Cumming of Altyre, is furnished with what seem uncouth paddle-shaped arms, that project from the head. All that I have given of the creature, however, will be found true to the actual type ; and that parts should have been omitted will surprise no one who remembers that many hundred belem-nites had been figured and described ere a spe-cimen turned up in which the horny prolonga-tion, with its inclosed ink bag, was found at-tached to the calcareous spindle ; and that even yet, after many thousand trilobites have been carefully examined, it remains a question with the oryctologist, whether this crustacean of the earliest periods was furnished with legs, or creep-ed on an abdominal foot, like the snail.

I owe to the kindness of Mr Robertson, In-verugie, the specimen figured in Plate V. fig. 7, containing shells of the only species yet disco-

---

Duff: see p. 130; **Lethen-bar**: see p. 129; **Gordon Cumming**: see p. 36; **spindle**: the belemnite skeleton is mostly spindle-shaped and rounded in cross-section, with a pointed end; **Robertson**: see p. 142.

vered in the Old Red Sandstone of Scotland.
They occur in the Lower Formation of the system, in a quarry near Kirkwall, in which the
specimen figured, with several others of the
same kind, was found by Mr Robertson in the
year 1834. In referring to this shell, page 99,
I have spoken of it as a delicate bivalve, much
resembling a *Venus ;* drawing my illustration
naturally enough, when describing the shell of
an ocean deposit, rather from among marine
than fluviatile testacea. I have since submitted
it to Mr Murchison, who has obligingly written
me, that he " can find no one to say more regarding it than that it is very like a *Cyclas.*"
He adds, however, that it must be an ocean
production notwithstanding, seeing that all its
cotemporaries in England, Scotland, and Russia, whether shells or fish, are unequivocally
marine.

With the exception of two of the figures in
Plate IX., the figures of the *Cephalaspis* and
the *Holoptychius*, and one of the sections in the
Frontispiece, section 2, all the prints of the
volume are originals. To Mr Daniel Alexander
of Edinburgh,—a gentleman who to the skill

---

shell: see p. 99; **Alexander**: see p. A99.

and taste of the superior artist, adds no small portion of the knowledge of the practical geologist,—I am indebted for several of the drawings : that of fig. 2 in Plate V., fig. 1 in Plate VI., fig. 2 in Plate VIII., and fig. 3 and 4 in Plate IX. I am indebted to another friend for fig. 1 in Plate VII. Whatever defects may be discovered in any of the others must be attributed to the untaught efforts of the writer, all unfamiliar hitherto with the pencil, and with by much too little leisure to acquaint himself with it now.

CONTENTS.

CHAPTER I.

CHAPTER II.

b

CHAPTER III.

CHAPTER IV.

CHAPTER V.

CHAPTER VI.

CHAPTER VII.

CHAPTER VIII.

CHAPTER IX.

CHAPTER XII.

CHAPTER XIII.

CHAPTER XIV.

EXPLANATIONS OF THE SECTIONS AND PLATES.

SECTION I.

Represents the Old Red System of Scotland from its upper beds of Yellow Quartzose Sandstone to its Great Conglomerate base. *a.* Quartzose Yellow Sandstone. *b.* Impure concretionary limestone inclosing masses of chert. *c.* Red and variegated sandstones and conglomerate. These three deposits constitute an upper formation of the system, characterized by its peculiar group of fossils (see Chapter IX.) *d.* Deposit of gray fissile sandstone which constitutes the middle formation of the system, characterized also by its peculiar organic group (see Chapter VIII.) *e.* Red and variegated sandstones undistinguishable often in their mineral character from the upper sandstones *c*, but in general less gritty, and containing fewer pebbles. *f.* Bituminous schists. *g.* Coarse gritty sandstone. *h.* Great Conglomerate. These four beds compose a lower formation of the system, more strikingly marked by its peculiar organisms than even the other two (see Chapters II. III. IV. and V.) In the section this lower formation is represented as we find it developed in Caithness and Orkney; in fig. 5 it is represented as developed in Cromarty, where, though the fossils are identical with those of the more northern localities, at least one of the deposits, *f*, is mineralogically different,— alternating beds of sandstone and clay, these last inclosing limestone nodules, taking the place of the bituminous schists.

SECTION II.

The Old Red System of England and Wales, as given in the general section of Mr Murchison, with the Silurian Rocks

---

general section: presumably based on Murchison 1839, part II, plate 31, fig. 1 (see p. 144), but if so, it is heavily generalised from the specific section in Murchison's plate.

beneath and the carboniferous limestone above. *i.* The point in the geological scale at which vertebrated existences first appear. The three Old Red Sandstone formations of this section correspond in their characteristic fossils with those of Scotland, but the proportions in which they are developed are widely different. The Tilestones seem a comparatively narrow stripe in the system in England: the answering formation in Scotland, *e, f, g, h,* is of such enormous thickness that it has been held by very superior geologists to contain three distinct formations,—*e,* the New Red Sandstone, *f,* a representative of the Coal Measures, and *g, h,* the Old Red Sandstone.

SECTION III.

Interesting case of extensive denudation from existing causes on the northern shore of the Moray Frith (see pages 205 and 206). The figures and letters which mark the various beds correspond with those of fig. 5, and of the following section. The " fish-bed," No. 1, represents what the reader will find described in pp. 232–236 as the " platform of sudden death."

SECTION IV.

Illustration of a fault in the Burn of Eathie, Cromartyshire (see pages 211 and 212).

Plate I. Fig. 1. Restoration of upper side of the elongated species of *Pterichthys* referred to in page 51. The specimens from which this figure has been chiefly taken are at present in the hands of Agassiz, from whom the writer has not yet received the creature's specific name. Fig. 2. *Pterichthys Milleri.* Fig. 3. Part of tail of elongated species, showing portions of the original covering of rhomboidal scales. Fig. 4. Tubercules of of *Pterichthys* magnified.

Plate II. Fig. 2. Restoration of under side of elongated *Pterichthys.* Fig. 1. A second specimen of *Pterichthys Milleri.* Fig. 3. Portion of wing, natural size.

Plate III. Fig. 1. *Coccosteus Cuspidatus.* Fig. 2. Impression of inner surface of large dorsal plate. Fig. 3. Abdominal lozenge shaped plate. Fig. 4. Portion of jaw with teeth.

---

superior geologists: presumably the initial reconnaissance by Sedgwick and Murchison 1829b, pp. 156–59.

Plate IV. Fig. 1. Restoration of *Osteolepis.* Fig. 2. Scales from the upper part of the body magnified. Fig. 3. Large defensive scale which runs laterally along all the single fins. Fig. 4. Under side of scale showing the attaching bar. Fig. 5. Enamelled and punctulated jaw of the creature. Fig. 6. Magnified portion of fin, showing the enamelled and punctulated rays. Fig. 7. Anterior ray of the dorsal fin.

Plate V. Fig. 1. *Dipterus.* This figure serves merely to show the place of the fins, and the general outline of the ichthyolite. All the specimens the writer has hitherto examined fail to show the minuter details. Fig. 2. *Glyptolepis.* Fig. 3. Single scale of the creature, showing its rustic style of ornament. Fig. 4. Scale with a nail-like attachment. Fig. 5. Under side of scale. Fig. 6. Magnified portion of fin. Fig. 7. Shells of the Old Red Sandstone.

Plate VI. Fig. 1. *Cheirolepis, Nov. Spec.* Fig. 2. Magnified scales. Fig. 3. Magnified portion of fin.

Plate VII. Fig. 1. *Cheiracanthus.* Fig. 2. Magnified Scales. Figs. 3, 4, 5, 6, 7, 8, Vegetable impressions of the Old Red Sandstone.

Plate VIII. Fig. 1. Ichthyolite not yet furnished with a name. Fig. 2. Ditto. Fig. 3. Magnified scales of fig. 1. Fig. 4. Spine of fig. 2, slightly magnified.

Plate IX. Fig. 1. *Cephalaspis Lyellii,* copied from Lyell's *Elements of Geology.* Fig. 2. *Holoptychius Nobilissimus,* copied on a greatly reduced scale from Murchison's *Silurian System.* Fig. 3. Scale of *Holoptychius,* natural size. Fig. 4. Tooth of ditto, also natural size. These last drawn from specimens in the collection of Mr Patrick Duff of Elgin.

ERRATA.

Page 15, line 8 from top, for *do* read *does*.

14, — 14 from bottom, for *Superior Oolite* read *Great Oolite*.

49, — 11 from top, for *less so* read *less deep*.

88, — 13 from bottom, for *Cheiracanthus* read *Cheirolepis*.

95, — 4 from top (in a few copies), for *Holoptychus* read *Holoptychius*.

147, — 12 from top, for *Crustaceon* read *Crustacean*.

171, — 9 from top, *delete* full stop and capital.

262, — 1 from top (in a few copies) *delete* full stop and capital.

NEW WALKS IN AN OLD FIELD;

OR

THE OLD RED SANDSTONE.

CHAPTER I.

The Working-man's True Policy.—His only Mode of acquiring Power.—The Exercise of the Faculties essential to Enjoyment.—No necessary Connection between Labour and Unhappiness.—Narrative.—Scenes in a Quarry.—The two Dead Birds.—Landscape.—Ripple Markings on a Sandstone Slab. — Boulder Stones. — Inferences derived from their water-worn Appearance. — Sea-coast Section. — My first-discovered Fossil.—Lias Deposit on the Shores of the Moray Frith.—Belemnite.—Result of the Experience of half a Lifetime of Toil.—Advantages of a Wandering Profession in connection with the Geology of a Country.—Geological Opportunities of the Stone-mason.—Design of the present Work.

My advice to young working men desirous to better their circumstances, and add to the amount of their enjoyment, is a very simple one. Do not seek happiness in what is misnamed pleasure; seek it rather in

A

Old Red Sandstone: the name for that particular 'system' or sequence of strata. Despite the name, and containing much actual sandstone, the Old Red Sandstone also includes other rocks, such as pebbly conglomerates, shales and limestones.

what is termed study. Keep your conscience clear,
your curiosity fresh, and embrace every opportunity
of cultivating your minds. You will gain nothing
by attending Chartist meetings. The fellows who
speak nonsense with fluency at these assemblies, and
deem their nonsense eloquence, are totally unable to
help either you or themselves, or, if they do succeed
in helping themselves, it will be all at your expense. Leave them to harangue unheeded, and set
yourselves to occupy your leisure hours in making
yourselves wiser men. Learn to make a right use
of your eyes;—the commonest things are worth looking at—even stones and weeds, and the most familiar
animals. Read good books, not forgetting the best
of all: there is more true philosophy in the Bible
than in every work of every sceptic that ever wrote ;
and we would be all miserable creatures without it,
and none more miserable than you. You are jealous
of the upper classes; and perhaps it is too true, that,
with some good, you have received much evil at their
hands. It must be confessed they have hitherto
been doing comparatively little for you, and a great
deal for themselves. But upper and lower classes
there must be, so long as the world lasts; and there
is only one way in which your jealousy of them
can be well directed. Do not let them get ahead
of you in intelligence. It would be alike unwise
and unjust to attempt casting them down to your
own level, and no class would suffer more in the
attempt than yourselves, for you would only be
clearing the way, at an immense expense of blood,
and under a tremendous pressure of misery, for an-

Chartist: of a then-active working-class movement whose aims included reform of
parliamentary representation with universal manhood suffrage. Widely regarded as
revolutionary, even insurrectionary. Miller, like many Whigs at the time, felt the vote
was best restricted to those who had a modicum of property and who were therefore
seen as more responsible (Taylor 2022a, pp. 84–85).

other and perhaps worse aristocracy, with some second Cromwell or Napoleon at their head. Society,
however, is in a state of continual flux : some in the
upper classes are from time to time going down, and
some of you from time to time mounting up to take
their places—always the more steady and intelligent
among you, remember; and if all your minds were
cultivated, not merely intellectually, but morally also,
you would find yourselves, as a body, in the possession
of a power which every charter in the world could
not confer upon you, and which all the tyranny or
injustice of the world could not withstand.

I intended, however, to speak rather of the pleasure
to be derived by even the humblest, in the pursuit of
knowledge, than of the power with which knowledge
in the masses is invariably accompanied. For it is
surely of greater importance that men should receive
accessions to their own happiness, than to the influence
which they exert over other men. There is none of
the intellectual, and none of the moral faculties, the
exercise of which does not lead to enjoyment; nay,
it is chiefly in the active employment of these that
all enjoyment consists; and hence it is that happiness
bears so little reference to station. It is a truth which
has been often told, but very little heeded or little
calculated upon, that though one nobleman may be
happier than another, and one labourer happier than
another, yet it cannot be at all premised of their respective orders, that the one is in any degree happier
than the other. Simple as the fact may seem, if uni-
versally recognized it would save a great deal of use-
less discontent, and a great deal of envy. Will my

humbler readers permit me at once to illustrate this
subject, and to introduce the chapters which follow,
by a piece of simple narrative? I wish to show them
how possible it is to enjoy much happiness in very
mean employments. Cowper tells us, that labour,
though the primal curse, " has been softened into
mercy;" and I think that, even had he not done so,
I would have found out the fact for myself.

It was twenty years last February since I set out
a little before sunrise to make my first acquaintance
with a life of labour and restraint, and I have rarely
had a heavier heart than on that morning. I was
but a slim, loose-jointed boy at the time—fond of
the pretty intangibilities of romance, and of dreaming
when broad awake; and, woful change! I was now
going to work at what Burns has instanced in his
" Twa Dogs," as one of the most disagreeable of all
employments—to work in a quarry. Bating the passing uneasinesses occasioned by a few gloomy antici-
pations, the portion of my life which had already gone
by had been happy beyond the common lot. I had
been a wanderer among rocks and woods—a reader
of curious books when I could get them—a gleaner
of old traditionary stories; and now I was going to
exchange all my day-dreams, and all my amusements,
for the kind of life in which men toil every day that
they may be enabled to eat, and eat every day that
they may be enabled to toil!

The quarry in which I wrought lay on the southern
shore of a noble inland bay, or frith rather, with a
little clear stream on the one side, and a thick fir
wood on the other. It had been opened in the Old

that morning: in February 1820: Miller 1995, p. 114; **softened …** : from William
Cowper's poem 'The Sofa', in *The Task*, book 1, where he reflects on the nature of work:
'see him sweating o'er his bread, / Before he eats it. 'Tis the primal curse / But softened
into mercy, made the pledge / Of cheerful days, and nights without a groan'; **Twa
Dogs**: (Two Dogs), Robert Burns's satirical poem in Ayrshire Scots comparing the
working masses with the idle rich, through the bemused eyes of the dogs Caesar and
Luath: 'they're fasht eneugh: / A cottar howkin in a sheugh, / Wi dirty stanes biggin a
dyke; baring a quarry / An sic like …' (they're bothered enough: a cottar digging in a

Red Sandstone of the district, and was overtopped by a huge bank of diluvial clay, which rose over it in some places to the height of nearly thirty feet, and which at this time was rent and shivered, wherever it presented an open front to the weather, by a recent frost. A heap of loose fragments, which had fallen from above, blocked up the face of the quarry, and my first employment was to clear them away. The friction of the shovel soon blistered my hands, but the pain was by no means very severe, and I wrought hard and willingly, that I might see how the huge strata below, which presented so firm and unbroken a frontage, were to be torn up and removed. Picks, and wedges, and levers were applied by my brother-workmen, and simple and rude as I had been accustomed to regard these implements, I found I had much to learn in the way of using them. They all proved inefficient, however, and the workmen had to bore into one of the inferior strata and employ gunpowder. The process was new to me, and I deemed it a highly amusing one : it had the merit, too, of being attended with some such degree of danger as a boating or rock excursion, and had thus an interest independent of its novelty. We had a few capital shots ; the fragments flew in every direction ; and an immense mass of the diluvium came toppling down, bearing with it two dead birds that, in a recent storm, had crept into one of the deeper fissures, to die in the shelter. I felt a new interest in examining them. The one was a pretty cock gold-finch, with its hood of vermilion, and its wings inlaid with the gold to which it owes its name, as unsoiled

---

pit, with dirty stones building a field wall; digging a quarry, and such like); **frith** (p. 4): Firth of Cromarty; **quarry** (p. 4): it lay about a hundred yards from the shore and was 'about thirty feet in height by eighty or a hundred in length', in the old sea-coast backing the raised beach: Miller 1838a, 1854a, pp. 148, 152; 1859, p. 44 (quoted); 1993, pp. 148, 152; 2022 [1858], p. 303; Gostwick 2013; Taylor and Morrison-Low 2017.

diluvium: superficial deposits originally thought to have been laid down by a major flood.

and smooth as if it had been preserved for a museum. The other, a somewhat rarer bird, of the woodpecker tribe, was variegated with light blue and a grayish yellow. I was engaged in admiring the poor little things, more disposed to be sentimental, perhaps, than if I had been ten years older, and thinking of the contrast between the warmth and jollity of their green summer haunts, and the cold and darkness of their last retreat, when I heard our employer bidding the workmen lay by their tools. I looked up, and saw the sun sinking behind the thick fir wood beside us, and the long dark shadows of the trees stretching downward towards the shore.

This was no very formidable beginning of the course of life I had so much dreaded. To be sure, my hands were a little sore, and I felt nearly as much fatigued as if I had been climbing among the rocks; but I had wrought and been useful, and had yet enjoyed the day fully as much as usual. It was no small matter, too, that the evening, converted, by a rare transmutation, into the delicious " blink of rest" which Burns so truthfully describes, was all my own. I was as light of heart next morning as any of my brother-workmen. There had been a smart frost during the night, and the rime lay white on the grass as we passed onward through the fields; but the sun rose in a clear atmosphere, and the day mellowed, as it advanced, into one of those delightful days of early spring, which give so pleasing an earnest of whatever is mild and genial in the better half of the year. All the workmen rested at mid day, and I went to enjoy my half-hour alone on a mossy knoll

---

bird: surely a nuthatch, presumably a vagrant as it is not native here, hence its strangeness to Miller; **'blink of rest'**: from Burns's 'Twa Dogs', on the honest satisfaction of toil as opposed to the idleness of the wealthy: 'tho fatigued wi close employment, / A blink o rest's a sweet enjoyment' (close = constant; blink = moment).

in the neighbouring wood, which commands through the trees a wide prospect of the bay and the opposite shore. There was not a wrinkle on the water, nor a cloud in the sky, and the branches were as moveless in the calm as if they had been traced on canvass. From a wooded promontory that stretched half-way across the frith there ascended a thin column of smoke. It rose straight as the line of a plummet for more than a thousand yards, and then, on reaching a thinner stratum of air, spread out equally on every side like the foliage of a stately tree. Ben Wevis rose to the west, white with the yet unwasted snows of winter, and as sharply defined in the clear atmosphere, as if all its sunny slopes and blue retiring hollows had been chiselled in marble. A line of snow ran along the opposite hills; all above was white, and all below was purple. They reminded me of the pretty French story, in which an old artist is described as tasking the ingenuity of his future son-in-law, by giving him as a subject for his pencil a flower-piece composed of only white flowers, the one half of them in their proper colour, the other half of a deep purple, and yet all perfectly natural; and how the young man resolved the riddle and gained his mistress, by introducing a transparent purple vase into the picture, and making the light pass through it on the flowers that were drooping over the edge. I returned to the quarry, convinced that a very exquisite pleasure may be a very cheap one, and that the busiest employments may afford leisure enough to enjoy it.

The gunpowder had loosened a large mass in one

Ben Wevis: Ben Wyvis; **pretty French story**: from the 1810 novel *Les Fleurs, ou les artistes*, in *La Botanique historique et littéraire* by the French aristocratic writer and educator Stéphanie de Genlis, perhaps in its 1811 translation (de Genlis 1811) or a review of it.

of the inferior strata, and our first employment, on resuming our labours, was to raise it from its bed. I assisted the other workmen in placing it on edge, and was much struck by the appearance of the platform on which it had rested. The entire surface was ridged and furrowed like a bank of sand that had been left by the tide an hour before. I could trace every bend and curvature, every cross hollow and counter ridge, of the corresponding phenomena; for the resemblance was no half resemblance—it was the thing itself; and I had observed it a hundred and a hundred times, when sailing my little schooner in the shallows left by the ebb. But what had become of the waves that had thus fretted the solid rock, or of what element had they been composed? I felt as completely at fault as Robinson Crusoe did on his discovering the print of the man's foot on the sand. The evening furnished me with still further cause of wonder. We raised another block in a different part of the quarry, and found that the area of a circular depression in the stratum below was broken and flawed in every direction, as if it had been the bottom of a pool recently dried up, which had shrunk and split in the hardening. Several large stones came rolling down from the diluvium in the course of the afternoon. They were of different qualities from the sandstone below, and from one another; and, what was more wonderful still, they were all rounded and water-worn, as if they had been tossed about in the sea or the bed of a river for hundreds of years. There could not surely be a more conclusive proof, that the bank which had enclosed them so long could

---

ridged and furrowed: this seemingly simple explanation was by no means obvious to all, as shown by a debate concerning similar finds in Sussex (Mantell 1831); Crusoe: alluding to the moment in *Robinson Crusoe*, Daniel Defoe's much-reprinted 'history' of 1719, in which Crusoe, marooned on an island and believing himself to be alone, is shocked to find the footprint of the man he later names 'Friday'.

not have been created on the rock on which it rested. No workman ever manufactures a half-worn article, and the stones were all half-worn! And if not
the bank, why then the sandstone underneath? I
was lost in conjecture, and found I had food enough
for thought that evening, without once thinking of
the unhappiness of a life of labour.

The immense masses of diluvium which we had
to clear away rendered the working of the quarry
laborious and expensive, and all the party quitted
it in a few days, to make trial of another that
seemed to promise better. The one we left is situated, as I have said, on the southern shore of an
inland bay—the bay of Cromarty; the one to which
we removed has been opened in a lofty wall of cliffs
that overhangs the northern shore of the Moray
Frith. I soon found I was to be no loser by the
change. Not the united labours of a thousand men
for more than a thousand years could have furnished a better section of the geology of the district
than this range of cliffs. It may be regarded as a
sort of chance dissection of the earth's crust. We see
in one place the primary rock, with its veins of granite and quartz—its dizzy precipices of gneiss, and
its huge masses of horneblend; we find the secondary rock in another, with its beds of sandstone and
shale—its spars, its clays, and its nodular limestones.
We discover the still little-known but highly interesting fossils of the Old Red Sandstone in one deposition—we find the beautifully preserved shells and lignites of the Lias in another. There are the remains
of two several creations at once before us. The shore

---

bay: Cromarty Bay, part of the Firth, just west of the town; **lofty wall**: exact location
untraced, but, in view of the need for a boat-landing to move the stone, presumably
the little bay beneath St Bennet's Well east of Navity farm: compare Miller 1854a,
pp. 152–53; **several**: distinct; **creations**: i.e. geological periods characterised by
different faunas and floras. See pp. A61–63.

too is heaped with rolled fragments of almost every variety of rock—basalts, ironstones, hyperstenes, porphyries, bituminous shales, and micaceous schists. In short, the young geologist, had he all Europe before him, could hardly choose for himself a better field. I had, however, no one to tell me so at the time, for geology had not yet travelled so far north; and so, without guide or vocabulary, I had to grope my way as I best might, and find out all its wonders for myself. But so slow was the process, and so much was I a seeker in the dark, that the facts contained in these few sentences were the patient gatherings of years.

In the course of the first day's employment, I picked up a nodular mass of blue limestone, and laid it open by a stroke of the hammer. Wonderful to relate, it contained inside a beautifully-finished piece of sculpture,—one of the volutes apparently of an Ionic capital; and not the far-famed walnut of the fairy tale, had I broken the shell and found the little dog lying within, could have surprised me more. Was there another such curiosity in the whole world? I broke open a few other nodules of similar appearance,—for they lay pretty thickly on the shore,—and found that there might. In one of these there were what seemed to be scales of fishes, and the impressions of a few minute bivalves, prettily striated; in the centre of another there was actually a piece of decayed wood. Of all nature's riddles these seemed to me to be at once the most interesting and the most difficult to expound. I treasured them carefully up, and was told by one of the workmen to

no one: in Cromarty the young Miller had neither local geologists nor books to help him – or even a **vocabulary**, the basic sorting and naming tool essential to any science. Sedgwick and Murchison had not yet visited the area; **volute**: spiral-scrolled ornament on a column capital of the Ionic order of Classical architecture; **nodules**: cast up here from offshore Jurassic deposits; **walnut**: in the German fairy tale 'Puddocky', then best known in Marie-Catherine d'Aulnoy's literary version of 1698 entitled 'The White Cat'.

whom I showed them, that there was a part of the shore about two miles farther to the west, where curiously-shaped stones, somewhat like the heads of boarding-pikes, were occasionally picked up ; and that in his father's days the country people called them thunder-bolts, and deemed them of sovereign efficacy in curing bewitched cattle. Our employer, on quitting the quarry for the building on which we were to be engaged, gave all the workmen a half-holiday. I employed it in visiting the place where the thunder-bolts had fallen so thickly, and found it a richer scene of wonder than I could have fancied in even my dreams.

What first attracted my notice was a detached group of low-lying skerries, wholly different in form and colour from the sandstone cliffs above, or the primary rocks a little farther to the west. I found they were composed of thin strata of limestone, alternating with thicker beds of a black slaty substance, which, as I ascertained in the course of the evening, burned with a powerful flame, and emitted a strong bituminous odour. The layers into which the beds readily separate are hardly an eighth part of an inch in thickness, and yet on every layer there are the impressions of thousands and tens of thousands of the various fossils peculiar to the Lias. We may turn over these wonderful leaves one after one, like the leaves of a herbarium, and find the pictorial records of a former creation in every page. Scallops, and gryphites, and ammonites of almost every variety peculiar to the formation, and at least two varieties of belemnite ; twigs of wood, leaves of plants, cones

---

to the west: at Eathie; **boarding-pikes**: for close combat on ships; **thunder-bolts ... cattle**: see Additional Notes; **herbarium**: collection of plants, in the sense of a bound album with a dried and pressed plant on each page.

of an extinct species of pine, bits of charcoal, and the scales of fishes; and, as if to render their pictorial appearance more striking, though the leaves of this interesting volume are of a deep black, most of the impressions are of a chalky whiteness. I was lost in admiration and astonishment, and found my very imagination paralyzed by an assemblage of wonders, that seemed to outrival in the fantastic and the extravagant, even its wildest conceptions. I passed on from ledge to ledge, like the traveller of the tale through the city of statues, and at length found one of the supposed aërolites I had come in quest of, firmly imbedded in a mass of shale. But I had skill enough to determine that it was other than what it had been deemed. A very near relative, who had been a sailor in his time on almost every ocean, and had visited almost every quarter of the globe, had brought home one of these meteoric stones with him from the coast of Java. It was of a cylindrical shape and vitreous texture, and it seemed to have parted in the middle when in a half-molten state, and to have united again, somewhat awry, ere it had cooled enough to have lost the adhesive quality. But there was nothing organic in its structure, whereas the stone I had now found was organised very curiously indeed. It was of a conical form and filamentary texture, the filaments radiating in straight lines from the centre to the circumference. Finely-marked veins like white threads ran transversely through these in its upper half to the point, while the space below was occupied by an internal cone, formed of plates that lay parallel to the base,

---

city: the City of Brass, a ghost city whose inhabitants died in a single act of divine judgement and are discovered *in situ* years later by travellers, in the narrative occupying Nights 567–78 of the Arabic story-collection *A Thousand and One Nights*: Lyons 2008, vol. II, pp. 519–46; **aërolites**: meteorites; **relative**: his father (Miller 1838a, 1995, p. 250); **meteoric**: possibly an oddly shaped concretion – but Java is a noted source area for tektites, droplets of molten rock splashed out by the impact of a large meteorite. Miller's description is, however, more typical of tektites from some other areas; **organic**: derived from a living organism; here, a fossil (in the modern sense of the word).

and which, like watch-glasses, were concave on the under side, and convex on the upper. I learned in time to call this stone a belemnite, and became acquainted with enough of its history to know that it once formed part of a variety of cuttle-fish, long since extinct.

My first year of labour came to a close, and I found that the amount of my happiness had not been less than in the last of my boyhood. My knowledge, too, had increased in more than the ratio of former seasons; and as I had acquired the skill of at least the common mechanic, I had fitted myself for independence. The additional experience of twenty years has not shown me that there is any necessary connection between a life of toil and a life of wretchedness; and when I have found good men anticipating a better and a happier time than either the present or the past, the conviction that in every period of the world's history the great bulk of mankind must pass their days in labour, has not in the least inclined me to scepticism.

My curiosity once fully awakened, remained awake, and my opportunities of gratifying it have been tolerably ample. I have been an explorer of caves and ravines, a loiterer along sea-shores, a climber among rocks, a labourer in quarries. My profession was a wandering one. I remember passing direct, on one occasion, from the wild western coast of Ross-shire, where the Old Red Sandstone leans at a high angle against the prevailing Quartz Rock of the district, to where, on the southern skirts of Mid-Lothian, the Mountain Limestone rises amid the coal. I have

---

a better and a happier time … scepticism: i.e. believing that their lives, and perhaps also their society, will improve over time; **passing direct**: in late 1823, Miller returned from work at Gairloch, in Wester Ross, to winter at Cromarty, before sailing to Edinburgh to work as a mason nearby (Taylor 2022a, pp. 29–32).

resided one season on a raised beach of the Moray
Frith.   I have spent the season immediately following amid the ancient granites and contorted schists
of the central Highlands.   In the north I have
laid open by thousands the shells and lignites of
the Oolite; in the south I have disinterred from
their matrices of stone or of shale the huge reeds
and tree ferns of the Carboniferous period.   I have
been taught by experience, too, how necessary an
acquaintance with the geology of both extremes of
the kingdom is to the right understanding of the
formations of either.   In the north there occurs a
vast gap in the scale.   The Lias leans unconformably
against the Old Red Sandstone; there is no Mountain Limestone, no Coal Measures, none of the New
Red Marls or Sandstones.   There are at least three
formations omitted.   But the upper portion of the
scale is well nigh complete.   In one locality we may
pass from the Lower to the Upper Lias, in another
from the Inferior to the Superior Oolite, and onward
to the Oxford Clay and the Coral Rag.   We may
explore in a third locality beds identical in their organisms with the Wealden of Sussex.   In a fourth
we find the flints and fossils of the Chalk.   The
lower part of the scale is also well nigh complete.
The Old Red Sandstone is amply developed in Moray, Caithness, and Ross, and the Grauwacke very
extensively in Banffshire.   But to acquaint one's self
with the three missing formations,—to complete one's
knowledge of the entire scale by filling up the hiatus,—
it is necessary to remove to the south. The geology
of the Lothians is the geology of at least two-thirds

---

raised beach: Cromarty burgh; **Highlands**: presumably at Gruids, near Lairg, where
he spent several autumn holidays with his aunt's family as a child: Miller 1854a; **north**:
i.e. of Scotland; **Oolite**: presumably Brora district: Sedgwick and Murchison 1829b;
unconformably: i.e. the Lias strata are differently oriented to those of the underlying
Old Red Sandstone. The eighteenth-century Scottish philosopher James Hutton
theorised that such unconformities indicated long gaps of time, allowing earth move-
ments to tilt the lower rock layers from their originally horizontal orientation before
the newer rocks were laid down (McIntyre and McKirdy 2012); **In one locality**: from

of the gap, and, perhaps, a little more ;—the geology of Arran wants only a few of the upper beds of the New Red Sandstone to fill it entirely.

One important truth I would fain press on the attention of my lowlier readers. There are few professions, however humble, that do not present their peculiar advantages of observation; there are none, I repeat, in which the exercise of the faculties do not lead to enjoyment. I advise the stone-mason, for instance, to acquaint himself with geology. Much of his time must be spent amid the rocks and quarries of widely separated localities. The bridge or harbour is no sooner completed in one district than he has to remove to where the gentleman's seat or farm-steading is to be erected in another; and so in the course of a few years he may pass over the whole geological scale, even when restricted to Scotland, from the Grauwacke of the Lammermuirs, to the Wealden of Moray or the Chalk-flints of Banffshire and Aberdeen, and this, too, with opportunities of observation at every stage, which can be shared with him by only the gentleman of fortune, who devotes his whole time to the study. Nay, in some respects, his advantages are superior to those of the amateur himself. The latter must often pronounce a formation unfossiliferous when, after the examination of at most a few days, he discovers in it nothing organic; and it will be found that half the mistakes of geologists have arisen from conclusions thus hastily formed. But the working man, whose employments have to be carried on in the same formation for months, per-

---

pp. 207 and 216, evidently Eathie to Navity near Cromarty; **another** (p. 14): Brora; **Superior** (p. 14): *recte* 'Great Oolite' (author's erratum, emended in 2nd edition); **third locality … Wealden** (p. 14): Linksfield near Elgin; **a fourth … flints and fossils of the Chalk** (p. 14): the Buchan Ridge gravels of Banffshire and Aberdeenshire. See p. 15; Sedgwick and Murchison 1829b, p. 158; Miller 2022 [1858], p. 234.

do not lead: *recte* 'does not lead' (author's erratum, emended in 2nd edition); **gentleman of fortune**: well off enough not to work for a living.

haps years together, enjoys better opportunities of
arriving at just decisions. There are, besides, a
thousand varieties of accident which lead to discovery,
—floods, storms, landslips, tides of unusual height,
ebbs of extraordinary fall; and the man who plies
his labour at all seasons in the open air has by much
the best chance of profiting by these. There are formations which yield their organisms slowly to the
discoverer, and the proofs which establish their place
in the geological scale even more tardily still. I was
acquainted with the Old Red Sandstone of Ross and
Cromarty for nearly ten years ere I had ascertained
that it is richly fossiliferous,—a discovery which, in
exploring this formation in those localities, some of
our first geologists had failed to anticipate: I was
acquainted with it for nearly ten years more ere I
could assign to its fossils their exact place in the
scale.

In the following chapters I shall confine my observations chiefly to this system and its organisms. To
none of the others, perhaps, excepting the Lias of the
north of Scotland, have I devoted an equal degree of
attention; nor is there a formation among them which,
up to the present time, has remained so much a *terra
incognita* to the geologist. The space on both sides
has been carefully explored to its upper and lower
boundary; the space between has been suffered to
remain well nigh a chasm. Should my facts regarding it—facts constituting the slow gatherings of years
—serve as stepping-stones laid across, until such
time as geologists of greater skill and more extended

---

accident: such as the overflowing millstream which revealed the Gamrie fish beds in
the late 1820s: Murchison 1828, p.363; **proofs**: of their age, from fossils and from field
relationships with known strata; **first geologists**: Murchison visited Cromarty in 1826
(but his main interest was in the Sutors and Eathie 'Lias'), and again with Adam
Sedgwick, apparently in 1827: Murchison 1827, 1828, p.356; Sedgwick and Murchison
1829b; **scale**: within the geological sequence, i.e. their relative age; *terra incognita*:
unknown land; **on both sides**: i.e. the Carboniferous above, well known from the
Midland Valley, and the Silurian below: Secord 1986.

research shall have bridged over the gap, I shall have completed half my design. Should the working man be encouraged by my modicum of success to improve his opportunities of observation, I shall have accomplished the whole of it. It cannot be too extensively known, that nature is vast and knowledge limited, and that no individual, however humble in place or acquirement, need despair of adding to the general fund.

B

CHAPTER II.

The Old Red Sandstone.—Till very lately its Existence as a
distinct Formation disputed.—Still little known.—Its great
Importance in the Geological Scale.—Illustration.—The
North of Scotland girdled by an immense Belt of Old Red
Sandstone.—Line of the Girdle along the Coast.—Marks of
vast Denudation.—Its Extent partially indicated by Hills
on the Western Coast of Ross-shire.—The System of great
Depth in the North of Scotland.—Difficulties in the way of
estimating the Thickness of Deposits.—Peculiar Formation
of Hill.—Illustrated by Ben Nevis.—Caution to the Geological Critic.—Lower Old Red Sandstone immensely developed in Caithness. — Sketch of the Geology of that
County.—Its strange Group of Fossils.—Their present place
of Sepulture.—Their ancient Habitat.—Agassiz.—Amazing Progress of Fossil Ichthyology during the last few
Years.—Its Nomenclature.—Learned Names repel unlearned Readers.—Not a great deal in them.

" The Old Red Sandstone," says a Scottish geologist, in a digest of some recent geological discoveries,
which appeared a short time ago in an Edinburgh
newspaper, " has been hitherto considered as remarkably barren of fossils." The remark is expressive of
a pretty general opinion among geologists of even the
present time, and I quote it on this account. Only
a few years have gone by since men of no low standing in the science disputed the very existence of this

---

geologist: Charles Maclaren (1782–1866), editor and co-proprietor of the *Scotsman*, the
Witness's rival Edinburgh newspaper, was presumably author of the digest (Anon. 1840b;
Matthew 2004); disputed: see Additional Notes.

formation—system rather, for it contains at least
three distinct formations; and, but for the influence
of one accomplished geologist, the celebrated author
of the *Silurian System*, it would have been probably degraded from its place in the scale altogether.
" You must inevitably give up the Old Red Sandstone," said an ingenious foreigner to Mr Murchison,
when on a visit to England about four years ago,
and whose celebrity among his own countrymen
rested chiefly on his researches in the more ancient
formations,—" you must inevitably give up the Old
Red Sandstone; it is a mere local deposit, a doubtful accumulation huddled up in a corner, and has
no type or representative abroad." " I would willingly give it up if nature would," was the reply; "but
it assuredly exists, and I cannot." In a recently-pub-
lished tabular exhibition of the geological scale by
a continental geologist, I could not distinguish the
system at all. There are some of our British geologists, too, who still regard it as a sort of debateable
tract, entitled to no independent status. They find,
in what they deem its upper beds, the fossils of the
Coal Measures, and the lower graduating apparently
into the Silurian System; and regard the whole as a
sort of common, which should be divided as proprietors used to divide commons in Scotland half a century ago, by giving a portion to each of the bordering
territories. Even the better-informed geologists, who
assign it its proper place as an independent formation
furnished with its own organisms, contrive to say all
they know regarding it in a very few paragraphs.
Lyell, in his admirable elementary work, published

---

celebrated author: Roderick Murchison, see Additional Notes; ingenious foreigner:
unidentified, but see Geikie 1875, vol. I, pp. 246–47. Miller's source was probably
Murchison himself, as, in the passage in the *Witness* equivalent to these lines ([Miller]
1840e, end of first paragraph), he mentions having corresponded with Murchison;
continental geologist: Alexandre Brongniart [1829]; Rudwick 2008, pp. 121–23; debateable tract: the Debateable Land, between the rivers Esk and Sark on the western border
between Scotland and England, and once claimed by both; Lyell: see Additional Notes.

only two years ago, devotes more than thirty pages to his description of the Coal Measures, and but two and a half to his notice of the Old Red Sandstone.*

* As the succinct notice of this distinguished geologist may serve as a sort of pocket-map to the reader in indicating the position of the system, its three great deposits, and its extent, I take the liberty of transferring it entire.

" OLD RED SANDSTONE.

" It was stated that the Carboniferous formation was surmounted by one called the ' New Red Sandstone,' and underlaid by another called the Old Red, which last was formerly merged in the Carboniferous system, but is now found to be distinguishable by its fossils. The Old Red Sandstone is of enormous thickness in Herefordshire, Worcestershire, Shropshire, and South Wales, where it is seen to crop out beneath the Coal-Measures, and to repose on the Silurian Rocks. In that region its thickness has been estimated by Mr Murchison at no less than 10,000 feet. It consists there of

" 1st, A quartzose conglomerate, passing downwards into chocolate-red and green sandstone and marl.

" 2d, Cornstone and marl (red and green argillaceous spotted marls, with irregular courses of impure concretionary limestone, provincially called Cornstone, mottled red and green; remains of fishes).

" 3d, Tilestone (finely laminated hard reddish or green micaceous or quartzose sandstones, which split into tiles; remains of mollusca and fishes)."

I have already observed that fossils are rare in marls and sandstones, in which the red oxide of iron prevails. In the Cornstone, however, of the counties above mentioned, fishes of the genera Cephalaspis and Onchus have been discovered. In the Tilestone also, Ichthyodorulites of the genus Onchus have been obtained, and a species of Dipterus, with mollusca of the genera Avicula, Arca, Cucullæa, Terebratula, Lingula, Turbo, Trochus, Turritella, Bellerophon, Orthoceras, and others.

By consulting geological maps the reader will perceive, that from Wales to the north of Scotland the Old Red Sandstone appears in patches, and often in large tracts. Many fishes

It will be found, however, that this hitherto ne-
glected system yields in importance to none of the
others, whether we take into account its amazing
depth, the great extent to which it is developed both
at home and abroad, the interesting links which it
furnishes in the zoological scale, or the vast period of
time which it represents. There are localities in
which the depth of the Old Red Sandstone fully
equals the elevation of Mount Etna over the level of
the sea, and in which it contains three distinct groupes
of organic remains, the one rising in beautiful progression over the other. Let the reader imagine a
digest of English history, complete from the times of
the invasion of Julius Cæsar to the reign of that Harold who was slain at Hastings, and from the times

have been found in it at Caithness, and various organic remains in the northern part of Fifeshire, where it crops out
from beneath the Coal formation, and spreads into the adjoining northern half of Forfarshire; forming, together with
trap, the Sidlaw Hills and valley of Strathmore. A large belt
of this formation skirts the northern borders of the Grampians, from the sea-coast at Stonehaven and the Frith of Tay,
to the opposite western coast of the Frith of Clyde. In Forfarshire, where, as in Herefordshire, it is many thousand feet
thick, it may be divided into three principal masses :—1st,
Red and mottled marls, cornstone, and sandstone; 2d, Conglomerate, often of vast thickness; 3d, Tilestones and paving
stone, highly micaceous, and containing a slight admixture of
carbonate of lime. In the uppermost of these divisions, but
chiefly in the lowest, the remains of fish have been found of
the genus named by M. Agassiz, Cephalaspis or buckler-
headed, from the extraordinary shield which covers the head,
and which has often been mistaken for that of a trilobite of
the division Asaphus. A gigantic species of fish of the genus
Gyrolepis has also been found by Dr Fleming in the Old Red
Sandstone of Fifeshire. (Lyell's *Elements*, pp. 452, 3, 4.)

Caesar ... Hastings: 55 BC to 1066 AD.

of Edward III. down to the present day, but bearing
no record of the Williams, the Henries, the Edwards,
the John, Stephen, and Richard, that reigned during the omitted period, or of the striking and important events by which their several reigns were distinguished. A chronicle thus mutilated and incomplete
would be no unapt representation of a geological history of the earth, in which the period of the Upper
Silurian would be connected with that of the Mountain Limestone, or of the limestone of Burdie House,
and the period of the Old Red Sandstone omitted.

The eastern and western coasts of Scotland which
lie to the north of the Friths of Forth and Clyde, together with the southern flank of the Grampians and
the northern coast of Sutherland and Caithness, appear to have been girdled at some early period by immense continuous beds of Old Red Sandstone. At
a still earlier time the girdle seems to have formed
an entire mantle which covered the enclosed tract
from side to side. The interior is composed of what,
after the elder geologists, I shall term primary rocks,
—porphyries, granites, gneisses, and micaceous schists;
and this central nucleus, as it now exists, seems set in
a sandstone frame. The southern bar of the frame
is still entire: it stretches along the Grampians from
Stonehaven to the Frith of Clyde. The northern bar
is also well nigh entire: it runs unbroken along the
whole northern coast of Caithness, and studs in three
several localities the northern coast of Sutherland,
leaving breaches of no very considerable extent between. On the east there are considerable gaps, as
along the shores of Aberdeenshire. The sandstone,

Edward III: ascended the throne in 1312 AD; **Burdie House**: Burdiehouse, now in south
Edinburgh; the Lower Carboniferous Burdiehouse Limestone was worked here, yielding important fossil fishes and plants (Andrews 1982a, pp. 13–15).

however, appears at Gamrie, in the county of Banff, in a line parallel to the coast, and, after another interruption, follows the course of the Moray Frith far into the interior of the great Caledonian valley, and then, running northward along the shores of Cromarty, Ross, and Sutherland, joins, after another brief interruption, the northern bar at Caithness. The western bar has also its breaches towards the south; but it stretches almost without interruption for about a hundred miles from the near neighbourhood of Cape Wraith to the southern extremity of Applecross; and though greatly disturbed and overflown by the traps of the inner Hebrides, it can be traced by occasional patches on towards the southern bar. It appears on the northern shore of Loch Alsh, on the eastern shore of Loch Eichart, on the southern shore of Loch Eil, on the coast and islands near Oban, and on the east coast of Arran. Detached hills and island-like patches of the same formation occur in several parts of the interior, far within the frame or girdle. It caps some of the higher summits in Sutherlandshire; it forms an oasis of sandstone among the primary districts of Strathspey; it rises on the northern shores of Lochness in an immense mass of conglomerate, based on a small-grained red granite, to a height of about three thousand feet over the level; and on the north-western coast of Ross-shire it forms three immense insulated hills, of at least no lower altitude, that rest unconformably on a base of gneiss.

There appear everywhere in connection with these patches and eminences, and with the surrounding

Gamrie: the parish containing the village of Gardenstown, which in Miller's day was also called Gamrie; **Cape Wraith**: Cape Wrath; **Loch Eichart**: Loch Eishort on Skye; Miller's account appears to be based on Sedgwick and Murchison 1829b, especially the map in plate 13.

girdle, marks of vast denudation.   I have often stood
fronting the three Ross-shire hills* at sunset in the
finer summer evenings, when the clear light threw the
shadows of their gigantic cone-like forms far over the
lower tract, and lighted up the lines of their horizontal
strata till they showed like courses of masonry in a pyramid. They seem at such times as if coloured by the
geologist, to distinguish them from the surrounding
tract, and from the base on which they rest as on a
common pedestal. The prevailing gneiss of the district
reflects a cold bluish hue, here and there speckled
with white, where the weathered and lichened crags
of intermingled Quartz-rock jut out on the hill-sides
from among the heath.   The three huge pyramids,
on the contrary, from the deep red of the stone, seem
flaming in purple.   There spreads all around a wild
and desolate landscape of broken and shattered hills,
separated by deep and gloomy ravines, that seem the
rents and fissures of a planet in ruins, and that speak
distinctly of a period of convulsion, when upheaving
fires from the abyss, and ocean-currents above, had
contended in sublime antagonism, the one slowly ele-
vating the entire tract, the other grinding it down and
sweeping it away. I entertain little doubt, that when
this loftier portion of Scotland, including the entire
Highlands, first presented its broad back over the
waves, the upper surface consisted exclusively, from
the one extremity to the other,—from Benlomond to
the Maidenpaps of Caithness,—of a continuous tract
of Old Red Sandstone, though, ere the land finally
emerged, the ocean-currents of ages had swept it

* Suil Veinn, Coul Beg, and Coul More.

---

coloured: as in a geological map or drawn cross-section (Figures 2, 11); **Benlomond to
the Maidenpaps**: from Ben Lomond, a mountain overlooking Loch Lomond, to
Maiden Pap, inland from Berriedale in Caithness: i.e. from one end of the Highland
massif to the other. In Sedgwick and Murchison 1829b, p. 128, Maiden Pap, Morven,
and other peaks are mentioned as 'a finely serrated mountain chain called the Maiden
Paps'; **Suil Veinn ...** : Suilven, Cùl Beag and Cùl Mor. Here and elsewhere, Miller
writes Gaelic in a roughly phonetic English transliteration.

away, all except in the lower and last-raised borders, and in the detached localities, where it still remains, as in the pyramidal hills of Western Ross-shire, to show the amazing depth to which it had once overlaid the inferior rocks. The Old Red Sandstone of Morvheim in Caithness overlooks all the primary hills of the district, from an elevation of three thousand five hundred feet.

The depth of the system, on both the eastern and western coasts of Scotland, is amazingly great—how great, I shall not venture to say. There are no calculations more doubtful than those of the geologist. The hill just instanced, Morvheim, is apparently composed from top to bottom of what in Scotland forms the lowest member of the system, a coarse conglomerate; and yet I have nowhere observed this inferior member, when I succeeded in finding a section of it directly vertical, more than a hundred yards in thickness—less than one-tenth the height of the hill. It would be well nigh as unsafe to infer, that the three thousand five hundred feet of altitude formed the real thickness of the conglomerate, as to infer that the thickness of the lead which covers the dome of St Paul's is equal to the height of the dome. It is always perilous to estimate the depth of a deposit by the height of a hill that seems externally composed of it, unless, indeed, like the pyramidal hills of Rossshire, it be unequivocally a hill dug out by denudation, as the sculptor digs his eminences out of the mass. In most of our hills the upheaving agency has been actively at work, and the space within is occupied by an immense nucleus of inferior rock,

---

Morvheim: Morven north of Helmsdale, whose modern Ordnance Survey height is 2313 feet (source: Sedgwick and Murchison 1829b, p. 128); St Paul's: St Paul's Cathedral, London; upheaving agency: on this view, many mountains were formed by injection of molten igneous rock heaving up wide domes of the country rock such as gneiss and mica-slate, with the injected granite sometimes breaking through the top. There might then be a further injection of porphyry which might itself break through the granite. Probably taken from Murchison (1828, pp. 355–56), Sedgwick and Murchison (1829a, pp. 35–36 n.), Anderson (1834, p. 194), and perhaps von Buch 1836. This seems to be a

around which the upper formation is wrapped like a caul, just as the vegetable mould or the diluvium wraps up this superior covering in turn. One of our best-known Scottish mountains—the gigantic Ben Nevis—furnishes an admirable illustration of this latter construction of hill. It is composed of three zones or rings of rock, the one rising over and out of the other, like the cases of an opera-glass drawn out. The lower zone is composed of gneiss and mica-slate,—the middle zone of granite,—the terminating zone of porphyry. The elevating power appears to have acted in the centre, as in the well-known case of Jorullo, in the neighbourhood of the city of Mexico, where a level tract four square miles in extent rose about the middle of the last century into a high dome of more than double the height of Arthur's Seat.* In the

---

* It is rarely that the geologist catches a hill in the act of forming, and hence the interest of this well-attested instance. From the period of the discovery of America to the middle of the last century, the plains of Jorullo had undergone no change of surface, and the seat of the present hill was covered by plantations of indigo and sugar-cane, when, in June 1759, hollow sounds were heard, and a succession of earthquakes continued for sixty days, to the great consternation of the inhabitants. After the cessation of these, and in a period of tranquillity, on the 28th and 29th September, a horrible subterranean noise was again heard, and a tract four square miles in extent rose up in the shape of a dome or bladder, to the height of 1670 feet above the original level of the plain. The affrighted Indians fled to the mountains; and from thence looking down on the phenomenon, saw flames issuing from the earth for miles around the newly-elevated hill, and the softened surface rising and falling like that of an agitated sea, and opening into numerous rents and fissures. Two brooks which had watered the plantations precipitated themselves into the burning chasms. The scene of this singular event

---

variant of the theory of 'craters of elevation' by Christian Leopold von Buch (1744–1853) (Oldroyd 1996a, 1996b, pp. 167–69).

vegetable mould: topsoil; **Jorullo**: El Jorullo, a volcano in central Mexico. Miller's source, actual or ultimate, was probably Humboldt 1811, vol. II, pp. 211–18; **Arthur's Seat**: the famous hill behind the Palace of Holyroodhouse in central Edinburgh, 823 feet high on modern survey.

formation of our Scottish mountain, the gneiss and mica-slate of the district seem to have been upheaved during the first period of Plutonic action in the locality, into a rounded hill of moderate altitude, but of huge base. The upheaving power continued to operate,—the gneiss and mica-slate gave way atop,— and out of this lower dome there arose a higher dome of granite, which, in an after and terminating period of the internal activity, gave way in turn to yet a third and last dome of porphyry. Now, had the elevating forces ceased to operate just ere the gneiss and mica-slate had given way, we would have known nothing of the interior nucleus of granite,— had they ceased just ere the granite had given way, we would have known nothing of the yet deeper nucleus of porphyry,—and yet the granite and the porphyry would assuredly have been there. Nor could any application of the measuring rule to the side of the hill have ascertained the thickness of its outer covering,—the gneiss and mica schist. The geologists of the school of Werner used to illustrate what we may term the anatomy of the earth, as seen through the spectacles of their system, by an onion

was visited by Humboldt about the beginning of the present century. At that period the volcanic agencies had become comparatively quiescent; the hill, however, retained its original altitude ; a number of smaller hills had sprung up around it ; and the traveller found the waters of the engulphed rivulets escaping at a high temperature from caverns charged with sulphurous vapours and carbonic acid gas. There were inhabitants of the country living at the time, who were more than twenty years older than the hill of Jorullo, and who had witnessed its rise.

Werner: Abraham Werner (1749–1817), lecturing at the Mining Academy in Freiberg in Saxony, interpreted many rocks, including in particular granite and basalt, as precipitates from a hot, mineral-rich primeval ocean; **Humboldt**: Alexander von Humboldt (1769–1859), German naturalist and geographer.

and its coats : they represented the globe as a central nucleus, encircled by concentric coverings, each
covering constituting a geological formation.   The
onion, through the introduction of a better school, has
become obsolete as an illustration ; but to restore it
again, though for another purpose, we have merely
to cut it through the middle, and turn downwards the
planes formed by the knife.   It then represents, with
its coats, two such hills as we describe,—hills such as
Ben Nevis, ere the granite had perforated the gneiss,
or the porphyry broken through the granite.

If it be thus unsafe, however, to calculate on the
depth of deposits by the altitude of hills, it is quite
as unsafe for the geologist who has studied a formation in one district, to set himself to criticise the calculations of a brother geologist, by whom it has been
studied in a different and widely-separated district.
A deposit in one locality may be found to possess
many times the thickness of the same deposit in another.   There are exposed beside the Northern and
Southern Sutors of Cromarty two nearly vertical sections of the coarse conglomerate bed, which forms, as
I have said, in the north of Scotland, the base of the
Old Red System, and which rises to so great an elevation in the mountain of Morvheim.   The sections are
little more than a mile apart; and yet while the thickness of this bed in the one does not exceed one hundred feet, that of the same bed in the other somewhat
exceeds two hundred feet.   More striking still,—under
the Northern Sutor, the entire Geology of Caithness,
with all its vast beds, and all its numerous fossils,
from the granitic rock of the Ord hill, the southern

---

better school: James Hutton (p. 14), in contrast to Werner, emphasised the igneous
origin of many rocks, notably basalt and granite, and the sedimentary origin of others;
but possibly Miller meant modern geology in general, including William Smith's
stratigraphy; turn downwards: each half onion has its cut face flat on the table; entire
Geology: Miller, like others, was assigning the Cromarty beds to the Lower Old Red
Sandstone; see Appendix 3.

boundary of the county, to the uppermost sandstones of Dunnet-head, its extreme northern corner, is exhibited in a vertical section not more than three hundred yards in extent. And yet so enormous is the depth of the deposit in Caithness, that it has been deemed by very superior geologists to represent three entire formations,—the Old Red System, by its unfossiliferous, arenaceous, and conglomerate beds,—the Carboniferous System, by its dark-coloured middle schists, abounding in bitumen and ichthyolites,—and the New Red Sandstone, by the mottled marls and mouldering sandstones that overlie the whole.* A slight sketch of the Geology of Caithness may not be deemed uninteresting. This county includes, in the state of greatest development anywhere yet known, that fossiliferous portion of the Old Red Sandstone which I purpose first to describe, and which will yet come to be generally regarded as an independent formation, as unequivocally characterised by its organic remains as the formations either above or below it.

The county of Sutherland stretches across the

* Dr Hibbert, whose researches among the limestones of Burdie House have been of such importance to Geology, was of this opinion. I find it also expressed in the admirable geological appendix affixed by the Messrs Anderson to their *Guide to the Highlands and Islands of Scotland.* " No beds of real coal," say these gentlemen, " have been discovered in Caithness; and it would thus appear that the middle schistose system of the county, containing the fossil fish, is in geological character and position intermediate between the Old and New Red Sandstone formations, but not identical with the Carboniferous Limestone or the true Coal Measures, although probably occupying the place of one or other of them." (P. 198.)

superior geologists: tactfully not naming Sedgwick and Murchison 1829b, pp.156–58, as followed by Anderson 1834; **schists**: here, shales rather than the modern meaning of a rock composed of metamorphosed shales; **organic remains**: fossil animals and plants; **Hibbert**: Dr Samuel Hibbert, later S. Hibbert-Ware (1782–1848), former soldier and keen geologist and antiquary instrumental in collecting the important Burdiehouse fossils: Hibbert 1835; Andrews 1982a; Sutton and Baigent 2004; **opinion**: source uncertain. Hibbert 1835, pp.231–33, covers only the area south of the Highlands; **appendix**: Anderson 1834, p.198, slightly edited. See also p.122 below.

island from the German to the Atlantic Ocean, and
presents throughout its entire extent,—except where a
narrow strip of the Oolitic formation runs along its
eastern coast, and a broken belt of Old Red Sandstone tips its capes and promontories on the west,—a
broken and tumultuous sea of primary hills. Scarce
any of our other Scottish counties are so exclusively
Highland, nor are there any of them in which the
precipices are more abrupt, the valleys more deep, the
rivers more rapid, or the mountains piled into more
fantastic groupes and masses. The traveller passes
into Caithness, and finds himself surrounded by scenery of an aspect so entirely dissimilar, that no examination of the rocks is necessary to convince him of a
geological difference of structure. An elevated and
uneven plain spreads around and before him, league
beyond league, in tame and unvaried uniformity,—its
many hollows darkened by morasses, over which the
intervening eminences rise in the form rather of low
moory swellings than of hills,—its coasts walled round
by cliffs of gigantic altitude, that elevate the district
at one huge stride from the level of the sea, and
skirted by vast stacks and columns of rock, that stand
out like the advanced picquets of the land amid the
ceaseless turmoil of the breakers. The district, as
shown on the map, presents nearly a triangular form,
—the Pentland Frith and the German Ocean describing two of its sidès, while the base is formed by the
line of boundary which separates it from the county
of Sutherland.

Now, in a geological point of view, this angle may
be regarded as a vast pyramid, rising perpendicularly

German ... Ocean: North Sea; picquets: sentries, outposts; triangular ... pyramid:
Miller here combines a two-dimensional geographical map of Caithness with a three-
dimensional geological model or 'pyramid' of its strata built upwards on the triangular
basis sketched out in the 'map'. His account appears to be based on Sedgwick and
Murchison 1829b. Miller was severely criticised by Robert Dick for bowing to ignorant
authority and oversimplifying the complexity of the Old Red in Caithness (Smiles 1878,
pp. 217–19); base: southern boundary line as seen from above or on a map.

from the basis furnished by the primary rocks of the
latter county, and presenting newer beds and strata
as we ascend, until we reach the apex. The line
from south to north in the angle,—from Morvheim to
Dunnet-head,—corresponds to the line of ascent from
the top to the bottom of the pyramid. The first bed,
reckoning from the base upwards,—the ground tier of
the masonry, if I may so speak,—is the great conglomerate. It runs along the line of boundary from sea
to sea,—from the Ord of Caithness on the east, to
Portskerry on the north; and rises, as it approaches
the primary hills of Sutherland, into a lofty mountain-
chain of bold and serrated outline, which attains its
greatest elevation in the hill of Morvheim. This
great conglomerate bed, the base of the system, is
represented in the Cromarty section, under the Northern Sutor, by a bed two hundred and fifteen feet in
thickness. The second tier of masonry in the pyramid, and which also runs in a nearly parallel line
from sea to sea, is composed mostly of a coarse red
and yellowish sandstone, with here and there beds
of pebbles inclosed, and here and there deposits of
green earth and red marl. It has its representative
in the Cromarty section in a bed of red and yellow
arenaceous stone, one hundred and fourteen feet six
inches in thickness. These two inferior beds possess
but one character;—they are composed of the same
materials, with merely this difference, that the rocks
which have been broken into pebbles for the construction of the one, have been ground into sand for
the composition of the other. Directly over them
the middle portion of the pyramid is occupied by an

---

basis: the foundation on which the pyramid of strata stands (not to be confused with
'base' on p. 30); **Portskerry**: Portskerra. Again, drawing heavily on Sedgwick and
Murchison 1829b.

enormous deposit of dark-coloured bituminous schist,
slightly micaceous, calcareous, or semi-calcareous,—
here and there interlaced with veins of carbonate of
lime,—here and there compact and highly siliceous,—
—and bearing in many places a mineralogical character difficult to be distinguished from that at one
time deemed peculiar to the harder grauwacke schists.
The Caithness flag-stones, so extensively employed in
paving the footways of our larger towns, are furnished by this immense middle tier or belt, and represent its general appearance. From its lowest to its
highest beds it is charged with fossil fish and obscure
vegetable impressions ; and we find it represented in
the Cromarty section by alternating bands of sandstones, stratified clays, and bituminous and nodular
limestones, which form altogether a bed three hundred and fifty-five feet in thickness ; nor does this
bed lack its organisms, animal and vegetable, generically identical with those of Caithness. The apex of
the pyramid is formed of red mouldering sandstones
and mottled marls, which exhibit their uppermost
strata high over the eddies of the Pentland Frith, in
the huge precipices of Dunnet-head, and which are
partially represented in the Cromarty section by an
unfossiliferous sandstone bed of unascertained thickness, but which can be traced for about eighty feet
from the upper limestones and stratified clays of the
middle member, until lost in overlying beds of sand
and shingle.

I am particular, at the risk, I am afraid, of being
tedious, in thus describing the Geology of this northern
county, and of the Cromarty section, which represents

and elucidates it. They illustrate more than the formations of two insulated districts; they represent also a vast period of time in the history of the globe. The pyramid with its three huge bars, its foundations of granitic rock, its base of red conglomerate, its central band of dark-coloured schist, and its lighter tinted apex of sandstone, is inscribed from bottom to top, like an Egyptian obelisk, with a historical record. The upper and lower sections treat of tempests and currents,— the middle is " written within and without" with wonderful narratives of animal life ; and yet the whole taken together comprises but an earlier portion of that chronicle of existences and events furnished by the Old Red Sandstone. It is, however, with this earlier portion that my acquaintance is most minute.

My first statement regarding it must be much the reverse of the borrowed one with which this chapter begins. *The fossils are remarkably numerous, and in a state of high preservation.* I have a hundred solid proofs by which to establish the truth of the assertion, within less than a yard of me. Half my closet walls are covered with the peculiar fossils of the Lower Old Red Sandstone; and certainly a stranger assemblage of forms have rarely been grouped together ;—creatures whose very type is lost,—fantastic and uncouth, and which puzzle the naturalist to assign them even their class ;—boat-like animals, furnished with oars and a rudder ;—fish plated over, like the tortoise, above and below, with a strong armour of bone, and furnished with but one solitary rudder-like fin ;—other fish, less equivocal in their form, but with the membranes of their fins thickly covered with scales ;—creatures

c

---

pyramid: see note for p. 30; **three huge bars**: here imagined in terms of their contrasting colours. Miller had described four 'tiers' on pp. 31–32, but the lower two were of the same 'character'; **'written within and without'**: Bible, Ezekiel 2: 9–10, from the prophet Ezekiel's vision, 'lo, a roll of a book was therein; and he spread it before me; and it was written within and without'; **very type is lost**: extinct; and more than this, that there are no longer any living relatives with those basic body plans, let alone any closely related genera or species.

bristling over with thorns; others glistening in an enamelled coat, as if beautifully japanned,—the tail, in every instance among the less equivocal shapes, formed not equally, as in existing fish, on each side the central vertebral bone, but chiefly on the lower side,— the bone sending out its diminished vertebræ to the extreme termination of the fin. All the forms testify of a remote antiquity,—of a period whose " fashions have passed away." The figures on a Chinese vase or an Egyptian obelisk are scarce more unlike what now exists in nature, than the fossils of the Lower Old Red Sandstone.

Geology, of all the sciences, addresses itself most powerfully to the imagination, and hence one main cause of the interest which it excites. Ere setting ourselves minutely to examine the peculiarities of these creatures, it were perhaps well that the reader should attempt realizing the *place* of their existence, and relatively the *time*,—not of course with regard to dates and eras, for the geologist has none to reckon by, but with respect to formations. They were the denizens of the same portion of the globe which we ourselves inhabit, regarded not as a tract of country, but as a piece of ocean crossed by the same geographical lines of latitude and longitude. Their present place of sepulture in some localities, had there been no denudation, would have been raised high over the tops of our loftiest hills,—at least a hundred feet over the conglomerates which form the summit of Morvheim,— and more than a thousand feet over the snow-capped Ben Wyvis. Geology has still greater wonders. I have seen belemnites of the Oolite,—comparatively a

---

japanned: lacquered; **'fashions have passed away'**: Bible, 1 Corinthians 7: 31, 'The fashion of this world passeth away'; **eras**: used here in the sense of a period of human history; the geochronological sense (such as 'Mesozoic era') was not widely used until later; **none to reckon by**: except for very recent events such as the eruption of Vesuvius in AD 79 which buried datable artefacts, geologists had no means of dating a rock's age in absolute terms of years passed; they could only give *relative* dates in relation to other strata; **belemnites**: probably from Malcolmson's Indian travels: p. 128.

modern formation,—which had been dug out of the
sides of the Himalaya mountains, seventeen thousand
feet over the level of the sea. But let us strive to
carry our minds back, not to the place of sepulture
of these creatures, high in the rocks,—though that I
shall afterwards attempt minutely to describe,—but to
the place in which they lived, long ere the sauroid
fishes of Burdie House had begun to exist, or the
coralines of the Mountain Limestone had spread out
their multitudinous arms in a sea gradually shallowing, and out of which the land had already partially
emerged.

A continuous ocean spreads over the space now
occupied by the British islands: in the tract covered
by the green fields and brown moors of our own
country, the bottom, for a hundred yards downwards,
is composed of the debris of rolled pebbles and coarse
sand intermingled, long since consolidated into the
lower member of the Old Red Sandstone; the upper
surface is composed of banks of sand, mud, and clay;
and the sea, swarming with animal life, flows over all.
My present object is to describe the inhabitants of
that sea.

Of these, the greater part yet discovered have been
named by Agassiz, the highest authority as an ichthyologist in Europe or the world, and in whom the
scarcely more celebrated Cuvier recognised a naturalist in every respect worthy to succeed him. The
comparative amount of the labours of these two great
men in fossil Ichthyology, and the amazing accelera-
tion which has taken place within the last few years
in the progress of geological science, are illustrated

---

coralines: 'coralline', in this context, means composed, associated with, or like coral.
Miller is using it as a noun to denote the organism; **Agassiz**: J. Louis R. Agassiz (1807–
73), Swiss palaeontologist and Professor of Natural History at Neuchâtel, and then the
main authority on fossil fishes, engaged in a systematic study of those fossils: Lurie 1988;
Andrews 1982a. See pp. A38–39, A46–51; **Cuvier**: Baron Georges Cuvier (1769–1832), the
great French anatomist and palaeontologist of the Muséum National d'Histoire Natur-
elle in Paris, had started a study of fossil fishes, but gave his materials to Agassiz in February 1832, so Miller must have been thinking on p. 36 of Agassiz's commencement of

together, and that very strikingly, by the following
interesting fact,—a fact derived directly from Agassiz
himself, and which must be new to the great bulk of
my readers.   When Cuvier closed his researches in
this department, he had named and described, for the
guidance of the geologist, ninety-two distinct species
of fossil fish ; nor was it then known that the entire
geological scale, from the Upper Tertiary to the Grauwacke inclusive, contained more.   Agassiz commenced his labours; and in a period of time little exceeding fourteen years, he has raised the number of species
from ninety-two to sixteen hundred.   And this number, great as it is, is receiving accessions almost every
day.   In his late visit to Scotland he found eleven
new species, and one new genus, in the collection of
Lady Cumming of Altyre, all from the upper beds of
that lower member of the Old Red Sandstone represented by the dark-coloured schists and inferior
sandstones of Caithness.   He found forty-two new
species more in a single collection in Ireland, furnished by the Mountain Limestone of Armagh.

Some of my humbler readers may possibly be re-
pelled by his names : they are, like all names in
science, unfamiliar in their aspect to mere English
readers, just because they are names not for England alone, but for England and the world.   I am
assured, however, that they are all composed of very
good Greek, and picturesquely descriptive of some
peculiarity in the fossils they designate.   One of his
ichthyolites with a thorn or spine in each fin, bears
the name of *Acanthodes*, or thorn-like , another,
with a similar mechanism of spines attached to the

work on *living* fishes in 1827 to get 14 years (Andrews 1982a, p. 7; Lurie 1988, pp. 28, 56).

late visit to Scotland: in 1840: see Appendix 2; **Lady Cumming**: Lady Eliza Gordon
Cumming (d.1842), of Altyre House near Forres, collected fossil fish from the area
(Andrews 1982a, 1983; Miller 2022 [1858], pp. 190–96; Trythall 2012; Appendix 2);
collection in Ireland: presumably of William Willoughby Cole, 3rd Earl of Enniskillen
(1807–86), at Florence Court, County Fermanagh: Andrews 1982a, p. 8; Davies 1970;
James 1986.

upper part of the body, and in which the pectoral or hand-fins are involved, has been designated the *Cheiracanthus,* or thorn-hand ; a third, covered with curiously-fretted scales, has been named the *Glypto-lepis,* or carved-scale ; and a fourth, roughened over with berry-like tubercules, that rise from strong os-seous plates, is known as the *Coccosteus,* or berry-on-bone. And such has been his principle of no-menclature. The name is a condensed description. But though all his names mean something, they can-not mean a great deal ; and as learned words repel unlearned readers, I shall just take the liberty of re-minding mine of the humbler class, that there is no legitimate connection between Geology and the dead languages. The existences of the Old Red Sandstone had lived for ages, and had been dead for myriads of ages, ere there was Greek enough in the world to fur-nish them with names. There is no working man, if he be a person of intelligence and information, how-ever unlearned in the vulgar acceptation of the phrase, who may not derive much pleasure and enlargement of idea from the study of Geology, and acquaint him-self as minutely with its truths as if possessed of all the learning of Bentley.

---

dead languages: Perhaps a response to William Cockburn, Dean of York, notorious amongst geologists for his ill-informed and often personal attacks on them and their science. Cockburn had just complained (1840, p. 9) that geologists used long words to 'throw a mist around the science, and frighten away the vulgar [i.e. the public] with unintelligible endecasyllables and cacophonous names'; **Bentley**: Richard Bentley (1662–1742), celebrated English Classical scholar who inaugurated the annual Boyle Lectures on religion and natural philosophy, initially envisaged as sermons wielding science against unbelief (Guerlac and Jacob 1969).

CHAPTER III.

Lamarck's Theory of Progression illustrated.—Class of Facts
which give Colour to it.—The Credulity of *Unbelief.*—M.
Maillet and his Fish-birds.—Gradation not Progress.—
Geological Argument.—The Present incomplete without
the Past.—Intermediate Links of Creation.—Organisms of
the Lower Old Red Sandstone.—The *Pterichthys.*—Its first
Discovery.—Mr Murchison's Decision regarding it.—Confirmed by that of Agassiz.—Description.—The several Varieties of the Fossil yet discovered.—Evidence of Violent
Death, in the Attitudes in which they are found.—The *Coccosteus* of the Lower Old Red.—Description.—Gradations
from Crustacea to Fishes.—Habits of the *Coccosteus.*—
Scarcely any Conception too Extravagant for Nature to
realize.

Mr Lyell's brilliant and popular work, *The
Principles of Geology,* must have introduced to the
knowledge of most of my readers the strange theories
of Lamarck. The ingenious foreigner, on the strength
of a few striking facts, which prove that, to a certain
extent, the instincts of species may be improved and
heightened, and their forms changed from a lower to
a higher degree of adaptation to their circumstances,
has concluded that there is a natural progress from
the inferior orders of being towards the superior; and
that the offspring of creatures low in the scale in the
present time, may hold a much higher place in it, and

Lamarck: Jean-Baptiste Lamarck (1744–1829), French naturalist, had argued, especially
in his *Philosophie zoologique* of 1809, that biological species did not become extinct, but
underwent transmutation to give a progressive, continuous development from simple
organisms to more complex ones. One of his proposed mechanisms was the fixation
and inheritance of physical changes acquired during an individual's adaptation to its
environment (Bowler 2009, pp. 86–95). The orang-utan and setter (next page) suggest
that Miller was drawing from Lyell 1830–33, vol. II, chapter 3, an influential critique of
Lamarckism which introduced it to many British readers.

belong to different and nobler species, a few thousand years hence. The descendants of the *ourang-outang*, for instance, may be employed in some future age in writing treatises on Geology, in which they shall have to describe the remains of the *quadrumana*, as belonging to an extinct order. Lamarck himself, when bearing home in triumph with him the skeleton of some huge salamander or crocodile of the Lias, might indulge, consistently with his theory, in the pleasing belief that he had possessed himself of the bones of his grandfather,—a grandfather removed, of course, to a remote degree of consanguinity, by the intervention of a few hundred thousand *great-greats*. Never yet was there a fancy so wild and extravagant but there have been men bold enough to dignify it with the name of philosophy, and ingenious enough to find reasons for the propriety of the name.

The setting-dog is *taught* to set; he squats down and points at the game; but the habit is an acquired one,—a mere trick of education. What, however, is merely acquired habit in the progenitor, is found to pass into instinct in the descendant: the puppy of the setting-dog squats down and sets *untaught*; the educational trick of the parent is mysteriously transmuted into an original principle in the offspring. The adaptation which takes place in the forms and constitution of plants and animals, when placed in circumstances different from their ordinary ones, is equally striking. The woody plant of a warmer climate, when transplanted into a colder, frequently exchanges its ligneous stem for a herbaceous one, as if in anticipation of the killing frosts of winter;

---

ourang-outang: orang-utan; **extinct**: i.e. the hypothetical descendants were no longer quadrumana, because they presumably each had two hands and two feet, not four hands; **setting-dog**: setter, as in Irish Setter; a type of gundog bred to 'set' to indicate hidden game birds, then by crouching but today by standing rigid.

and, dying to the ground at the close of autumn, shoots
up again in spring.   The dog, transported from a
temperate into a frigid region, exchanges his covering
of hair for a covering of wool: when brought back
again to his former habitat, the wool is displaced
by the original hair.   And hence, and from similar
instances, the derivation of an argument, good so far
as it goes, for changes in adaptation to altered circumstances of the organization of plants and animals, and
for the improvability of instinct.   But it is easy driving
a principle too far.   The elasticity of a common bow,
and the strength of an ordinary arm, are fully adequate to the transmission of an arrow from one point
of space to another point a hundred yards removed ;
but he would be a philosopher worth looking at who
would assert that they were equally adequate for the
transmission of the same arrow from points removed,
not by a hundred yards, but by a hundred miles.
And such, but still more glaring, has been the error
of Lamarck.   He has argued on this principle of
improvement and adaptation, which, carry it as far
as we rationally may, still leaves the vegetable a
vegetable, and the dog a dog, that in the vast course
of ages, inferior have risen into superior natures,
and lower into higher races ; that molluscs and zoophites have passed into fish and reptiles, and fish
and reptiles into birds and quadrupeds ; that un-
formed gelatinous bodies, with an organization
scarcely traceable, have been metamorphosed into
oaks and cedars ; and that monkeys and apes have
been transformed into human creatures, capable of
understanding and admiring the theories of Lamarck.

Assuredly, there is no lack of faith among infidels;
their "vaulting" credulity o'erleaps revelation, and
"falls on the other side." One of the first geological
works I ever read was a philosophical romance, entitled *Teliamed*, by a Mon. Maillet, an ingenious French-
man of the days of Louis XV. This Maillet was by
much too great a philosopher to credit the scriptural
account of Noah's flood, and yet he could believe,
like Lamarck, that the whole family of birds had
existed at one time as fishes, which, on being thrown
ashore by the waves, had got feathers by accident;
and that men themselves are but the descendants of
a tribe of sea-monsters, who, tiring of their proper element, crawled up the beach one sunny morning, and,
taking a fancy to the land, forgot to return.*

* Few men could describe better than Maillet. His extravagancies are as amusing as those of a fairy tale, and quite
as extreme. Take the following extract as an instance :—
" Winged or flying fish, stimulated by the desire of prey, or
the fear of death, or pushed near the shore by the billows,
have fallen among reeds or herbage, whence it was not possible for them to resume their flight to the sea, by means of
which they had contracted their first facility of flying. Then
their fins, being no longer bathed in the sea-water, were split,
and became warped by their dryness. While they found
among the reeds and herbage among which they fell, any ali-
ments to support them, the vessels of their fins being separated, were lengthened and clothed with beards, or, to speak
more justly, the membranes which before kept them adherent
to each other, were metamorphosed. The beard formed of
these warped membranes was lengthened. The skin of these
animals was insensibly covered with a down of the same colour with the skin, and this down gradually increased. The
little wings they had under their belly, and which, like their
wings, helped them to walk in the sea, became feet, and served
them to walk on land. There were also other small changes

'**vaulting ... side**': adapted from Macbeth's soliloquy in Shakespeare's *Macbeth*, Act 1,
Scene 7, 'vaulting ambition, which o'erlaps itself, / And falls on th'other'; **revelation**: revealed religion, here Christianity; **Noah's flood**: then no longer taken as a global inundation by reputable geologists; ***Teliamed***: *Telliamed*, by the French diplomat Benoît de
Maillet (1656–1738), published in 1748. Miller used the English edition (Maillet 1750,
pp. 223–24). Its hypotheses about the evolution of life and emergence of landforms from
the sea over billions of years are presented in the voice of an 'Indian philosopher' whose
name was 'de Maillet' spelt backwards (Carozzi 1968, 1969). See p. A67.

" How easy," says this fanciful writer, " is it to conceive the change of a winged fish, flying at times through the water, at times through the air, into a bird flying always through the air!" It is a law of nature that the chain of being, from the lowest to the highest form of life, should be, in some degree, a continuous chain; that the various classes of existence should shade into one another, so that it often proves a matter of no little difficulty to point out the exact line of demarcation where one class or family ends and another class or family begins. The naturalist passes from the vegetable to the animal tribes, scarcely aware, amid the perplexing forms of intermediate existence, at what point he quits the precincts of the one to enter on those of the other. All the animal families have, in like manner, their connecting links; and it is chiefly out of these that writers such as Lamarck and Maillet construct their system. They confound gradation with progress. Geoffrey

in their figure. The beak and neck of some were lengthened, and those of others shortened. The conformity, however, of the first figure subsists in the whole, and it will be always easy to know it. Examine all the species of fowls, large and small, even those of the Indies, those which are tufted or not, those whose feathers are reversed, such as we see at Damietta, that is to say, whose plumage runs from the tail to the head, and you will find species quite similar, scaly or without scales. All species of parrots, whose plumages are so different, the rarest and the most singular-marked birds, are, conformable to fact, painted like them with black, brown, gray, yellow, green, red, violet-colour, and those of gold and azure; and all this precisely in the same parts where the plumages of those birds are diversified in so curious a manner." (*Teli-amed*, p. 224, ed. 1750.)

---

chain of being ... links: see pp. A67–68; **Geoffrey Hudson**: Jeffrey Hudson (1619–c.1682), 'Queen's dwarf' to Henrietta Maria, consort of the Stuart king Charles I.

Hudson was a very short man, and Goliath of Gath a very tall one, and the gradations of the human stature lie between. But gradation is not progress; and though we find full-grown men of five feet, five feet six inches, six feet, and six feet and a half, the fact gives us no earnest whatever that the race is rising in stature, and that at some future period the average height of the human family will be somewhat between ten and eleven feet. And equally unsolid is the argument that from a principle of gradation in races would deduce a principle of progress in races. The tall man of six feet need entertain quite as little hope of rising into eleven feet, as the short man of five; nor has the fish that occasionally flies any better chance of passing into a bird, than the fish that only swims.

Geology abounds with creatures of the intermediate class: there are none of its links more numerous than its connecting links; and hence its interest, as a field of speculation, to the assertors of the transmutation of races. But there is a fatal incompleteness in the evidence, that destroys its character as such. It supplies in abundance those links of generic connection which, as it were, marry together dissimilar races; but it furnishes no genealogical link to show that the existences of one race derive their lineage from the existences of another. The scene shifts as we pass from formation to formation; we are introduced in each to a new *dramatis personæ ;* and there exist no such proofs of their being at once different and yet the same, as those produced in the *Winter's Tale* to show that the grown shepherdess of the one scene is identical with the exposed infant of the scene that

---

Goliath of Gath: giant champion of the Philistines slain by David: Bible, 1 Samuel 17; **Geology ... links**: i.e. the points (links) in the chain of being which are seen in the fossil record are most often intermediate (connecting links) between one present-day creature and another; **transmutation of races**: evolution, in the modern sense; *Winter's Tale*: Shakespeare's play, featuring a baby abandoned in a forest but rescued by a shepherd at the end of Act 3. In Act 4 she appears as a grown woman, 16 years having passed between the two scenes without being represented in the action (on this allusion see Geric 2017, pp. 47–49).

went before. Nay, the reverse is well nigh as strikingly the case, as if the grown shepherdess had been
introduced into the earlier scenes of the drama, and
the child into its concluding scenes.

The argument is a very simple one. Of all the
vertebrata, fishes rank lowest, and in geological history
appear first. We find their remains in the Upper
Ludlow Rocks, in the Lower, Middle, and Upper
Old Red Sandstone, in the Mountain Limestone, and
in the Coal Measures: we find them also in the
Magnesian Limestone; and in the latter formation
the first reptiles appear. Fishes seem to have been
the master existences of five succeeding formations,
ere the age of reptiles began. Now fishes differ very
much among themselves: some rank nearly as low
as worms, some nearly as high as reptiles; and if
fish could have risen into reptiles, and reptiles into
mammalia, we would necessarily expect to find lower
orders of fish passing into higher, and taking precedence of the higher in their appearance in point of
time, just as in the *Winter's Tale* we see the infant preceding the adult. If such be not the case,—
if fish made their first appearance, not in their least
perfect, but in their most perfect state,—not in their
nearest approximation to the worm, but in their nearest approximation to the reptile,—there is no room
for progression, and the argument falls. Now, it is a
geological fact, that it is fish of the higher orders that
appear first on the stage, and that they are found to
occupy exactly the same level during the vast period
represented by five succeeding formations. There is
no progression. If fish rose into reptiles, it must have

---

progression: evolution (i.e. transmutation of species) in a specific direction, usually
from simple to complex, or from microbes to humans; **Milton's sublime figure** (p. 45):
John Milton (1608–74), English Puritan poet, in *Areopagitica* (1644), an eloquent
protest against press censorship. Milton presents Truth as fragmented and unattainable
in its complete form, and partially accessible through reasoned debate and controversy:
'the sad friends of Truth … imitating the careful search that Isis made for the mangled
body of Osiris, went up and down gathering up limb by limb still as they could find
them. We have not yet found them all, Lords and Commons, nor ever shall do, till her

been by sudden transformation ;—it must have been as if a man who had stood still for half a life-time should bestir himself all at once, and take seven leagues at a stride. There is no getting rid of miracle in the case,—there is no alternative between creation and metamorphosis. The infidel substitutes progression for Deity ; Geology robs him of his god.

But no man who enters the geological field in quest of the wonderful, need pass in pursuit of his object from the true to the fictitious. Does the reader remember how, in Milton's sublime figure, the body of Truth is represented as hewn in pieces, and her limbs scattered over distant regions, and how her friends and disciples have to go wandering all over the world in quest of them ? There is surely something very wonderful in the fact, that in uniting the links of the chain of creation into an unbroken whole, we have in like manner to seek for them all along the scale of the geologist ;—some we discover among the tribes first annihilated,—some among the tribes that perished at a later period,—some among the existences of the passing time. We find the present incomplete without the past,—the recent without the extinct. There are marvellous analogies which pervade the scheme of Providence, and unite, as it were, its lower with its higher parts. The perfection of the works of Deity is a perfection entire in its components, and yet these are not contemporaneous, but successive : it is a perfection which includes the dead as well as the living, and bears relation in its completeness, not to time, but to eternity.

We find the organisms of the Old Red Sandstone

Master's [Christ] second coming; He shall bring together every joint and member, and shall mould them into an immortal feature of loveliness and perfection'; **scheme**: the chain of being (see p. 42), its perfection manifest only when taking into account all past and future created beings, not just the present world.

supplying an important link, or rather series of links, in the ichthyological scale, which are wanting in the present creation, and the absence of which evidently occasions a wide gap between the two grand divisions or series of fishes,—the bony and the cartilaginous. Of this, however, more anon. Of all the organisms of the system, one of the most extraordinary, and in which Lamarck would have most delighted, is the *Ptericthys*, or winged fish, an ichthyolite which the writer had the pleasure of introducing to the acquaintance of geologists nearly three years ago, but which he first laid open to the light about seven years earlier. Had Lamarck been the discoverer, he would unquestionably have held that he had caught a fish almost in the act of wishing itself into a bird. There are wings which want only feathers, a body which seems to have been as well adapted for passing through the air as the water, and a tail by which to steer. And yet there are none of the fossils of the Old Red Sandstone which less resemble any thing that now exists than its *Ptericthys*. I fain wish I could communicate to the reader the feeling with which I contemplated my first-found specimen. It opened with a single blow of the hammer; and there, on a ground of light-coloured limestone, lay the effigy of a creature fashioned apparently out of jet, with a body covered with plates, two powerful-looking arms articulated at the shoulders, a head as entirely lost in the trunk as that of the ray or the sunfish, and a long angular tail. My first-formed idea regarding it was, that I had discovered a connecting link between the tortoise and the fish; the body much

---

present creation: i.e. living animals and plants; **Ptericthys**: a then-common misspelling of 'ichthy-' compounds; emended to *Pterichthys* in the 3rd edition; **introducing to … geologists**: see Additional Notes.

PLATE I.

Ptericthys

Fig. 1.

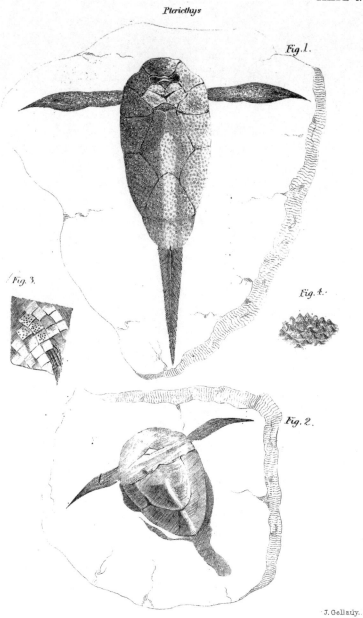

Fig. 3.

Fig. 4.

Fig. 2.

J. Gellatly.

Published by John Johnstone, Hunter Square, Edinburgh.

resembles that of a small turtle; and why, I asked,
if one formation gives us sauroid fishes, may not another give us chelonian ones? or if in the Lias we
find the body of the lizard mounted on the paddles
of the whale, why not find in the Old Red Sandstone
the body of the tortoise mounted in a somewhat similar manner? The idea originated in error, but as it
was an error which not many naturalists could have
corrected at the time, it may be deemed an excusable one, more especially by such of my readers
as may have seen well-preserved specimens of the
creature, or who examine the subjoined prints, No.
I. and II. I submitted some of my specimens to
Mr Murchison, at a time when that gentleman was
engaged among the fossils of the Silurian System,
and employed on his great work, which has so
largely served to extend geological knowledge regarding those earlier periods in which animal life
first began. He was much interested in the discovery : it furnished the geologist with additional data
by which to regulate and construct his calculations,
and added a new and very singular link to the chain
of existence in its relation to human knowledge.
Deferring to Agassiz, as the highest authority, he yet
anticipated the decision of that naturalist regarding
it in almost every particular. I had inquired, under
the influence of my first impression, whether it might
not be considered as a sort of intermediate existence
between the fish and the chelonian. He stated, in reply, that he could not deem it referrible to any family
of reptiles; that if not a fish, it approached more closely
to the crustacea than to any other class; and that he

---

sauroid fishes: such as the giant predatory forms of the Burdiehouse Limestone, which
were reminiscent of saurian reptiles such as crocodiles with their ferocious teeth and
large size. Early finds of teeth were initially mistaken for those of reptiles (Hibbert 1835;
Andrews 1982a, pp. 13–18); **body of the lizard mounted on the paddles of the whale**:
almost certainly the ichthyosaur as described by Buckland (1836, vol. I, pp. 183–85);
submitted: in mid-1838; **great work**: *The Silurian System*, Murchison 1839.

had little doubt Agassiz would pronounce it to be an ichthyolite of that ancient order to which the *Cephalaspis* belongs, and which seems to have formed a connecting link between crustacea and fishes.* The specimens submitted to Mr Murchison were forwarded to Agassiz. They were much more imperfect than some which I have since disinterred; and to restore the entire animal from them would require powers such as those possessed by Cuvier in the past age, and by the naturalist of Neufchatel in the present. Broken as they were, however, Agassiz at once decided from them that the creature must have been a fish.

* The aborigines of South America deemed it wonderful that the Europeans who first visited them should, without previous concert, agree in reading after the same manner the same scrap of manuscript, and in deriving the same piece of information from it. The writer experienced on this occasion a somewhat similar feeling. His specimens seemed written in a character cramp enough to suggest those doubts regarding original meaning which lead to various readings, but the geologist and the naturalist agreed in perusing them after exactly the same fashion, the one in London, the other in Neufchatel. Such instances give confidence in the findings of science. The decision of Mr Murchison I subjoin in his own words: his numbers refer to various specimens of *Pterichthys* :—" As to your fossils 1, 2, 3, we know nothing of them here (London), except that they remind me of the occipital fragments of some of the Caithness fishes. I do not conceive they can be referrible to any reptile ; for, if not fishes, they more closely approach to crustaceans than to any other class. I conceive, however, that Agassiz will pronounce them to be fishes which, together with the curious genus *Cephalaspis* of the Old Red Sandstone, form the connecting links between crustaceans and fishes. Your specimens remind one in several respects of the *Cephalaspis.*"

Murchison and footnote: letter to Miller of 23 June 1838, HMLB item 194, printed in Bayne 1871, vol. II, pp. 148–50 (but, wrongly, with 'clearly' for **closely**), and, partly, in Geikie 1875, part I, pp. 259–60.

I have placed one of the specimens before me. Imagine the figure of a man rudely drawn in black on a gray ground, the head cut off by the shoulders, the arms spread at full, as in the attitude of swimming, the body rather long than otherwise, and narrowing from the chest downwards, one of the legs cut away at the hip joint, and the other, as if to preserve the balance, placed directly under the centre of the figure, which it seems to support. Such, at a first glance, is the appearance of the fossil. The body was of very considerable depth, perhaps little less so proportionally from back to breast than the body of the tortoise; the under part was flat, the upper rose towards the centre into a roof-like ridge, and both under and upper were covered with a strong armour of bony plates, which, resembling more the plates of the tortoise than those of the crustacean, received their accessions of growth at the edges or sutures. The plates on the under side are divided by two lines of suture, which run, the one longitudinally through the centre of the body, the other transversely, also through the centre of it ; and they would cut one another at right angles, were there not a lozenge-shaped plate inserted at the point where they would otherwise meet. There are thus five plates on the lower or belly part of the animal. They are all thickly tuberculated outside with wart-like prominences; the inner present appearances indicative of a bony structure. The plates on the upper side are more numerous and more difficult to describe, just as it would be difficult to describe the forms of the various stones which compose the ribbed and pointed roof of a Gothic

D

little less so: should read 'little less deep' (author's erratum, emended in 2nd edition); accessions of growth: *Pterichthys's* plates grew by adding new bone to the edge, leaving characteristic concentric lines, as in the shell of a tortoise. By contrast, the crustacean shell does not grow and is periodically moulted, the animal expanding while the new shell is still soft.

cathedral, the arched ridge or hump of the back re-
quiring, in a somewhat similar way, a peculiar form
and arrangement of plates. The apex of the ridge
is covered by a strong hexagonal plate, fitted upon it
like a cap or helmet, and which nearly corresponds in
place to the flat central plate of the under side.
There runs around it a border of variously-formed
plates, that diminish in size and increase in number
towards the head, and which are separated like the
pieces of a dissected map, by deep sutures. They all
present the tuberculated surface. The eyes are placed
in front, on a prominence much lower than the roof-
like ridge of the back; the mouth seems to have
opened, as in many fishes, in the edge of the creature's snout, where a line running along the back
would bisect a line running along the belly, but this
part is less perfectly shown by my specimens than any
other. The two arms or paddles are placed so far
forward as to give the body a disproportionate and
decapitated appearance. From the shoulder to the
elbow, if I may employ the terms, there is a swelling
muscular appearance, as in the human arm ; the part
below is flattened so as to resemble the blade of an
oar, and it terminates in a strong sharp point. The
tail—the one leg on which, as exhibited in one of
my specimens, the creature seems to stand—is of
considerable length, more than equal to a third of the
entire figure, and of an angular form, the base representing the part attached to the body, and the apex
its termination. It was covered with small tuberculated angular plates like scales, and where the internal
structure is shown, there are appearances of a verte-

PLATE II.

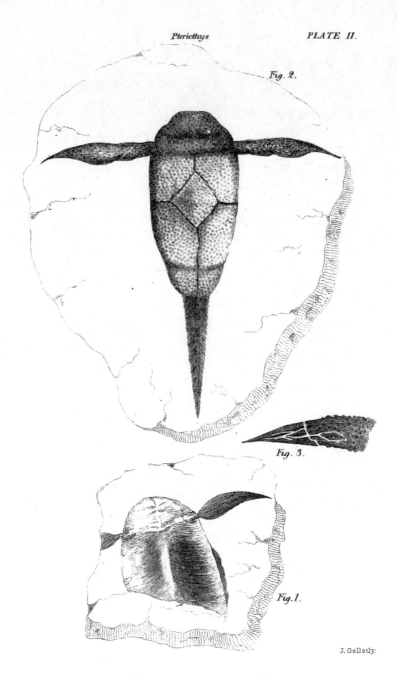

Fig. 2.

Fig. 3.

Fig. 1.

J. Gellatly.

Published by John Johnstone, Hunter Square. Edinburgh.

brated bone, with rib-like processes standing out at a
sharp angle. The ichthyolite in my larger specimens
does not much exceed seven inches in length, and I
despatched one to Agassiz rather more than two years
ago, whose extreme length did not exceed an inch.
Such is a brief, and, I am afraid, imperfect sketch of
a creature whose very type seems no longer to exist.
But for the purposes of the geologist the descriptions
of the graver far exceed those of the pen, and the accompanying prints will serve to supply all that may be
found wanting in the text. Fig. 1 in Plate I. and
fig. 2 in Plate II. are both restorations, the first of the
upper, and the second of the under part of the creature. It may, however, encourage the confidence of
the naturalist, who for the first time looks upon forms
so strange, to be informed that Plate I., with its two
figures, was submitted to Agassiz during his recent
brief stay in Edinburgh, and that he as readily recognised in it the species of the two varieties which it
exhibits, as he had previously recognised the species
of the originals in the limestone.

Agassiz, in the course of his late visit to Scotland,
found six species of the *Ptericthys*,—three of these, and
the wings of a fourth, in the collection of the writer.
The differences by which they are distinguished may
be marked by even an unpractised eye, especially in
the form of the bodies and wings. Some are of a
fuller, some of a more elongated form ; in some the
body resembles a heraldic shield, of nearly the ordinary
shape and proportions ; in others the shield stretches
into a form not very unlike that of a Norway skiff,
from the midships forward. In some of the varieties,

very type: see p. 33; **stay … visit**: late October 1840 (Appendix 2); **two varieties**: not
'varieties' in today's technical sense, but just different kinds: *Pterichthys oblongus* and
P. milleri (see Appendix 4); **six species**: in Agassiz's taxonomy. *Pterichthys milleri*,
P. oblongus, and *P. testudinarius* at least were in Miller's collection, and the others were
P. latus, *P. cancriformis* and *P. hydrophilus*: Agassiz 1844–45; Andrews 1982a, esp. pp. 29,
42–43; **Norway skiff**: traditional Norse clinker-built rowing boat, also exported to
Scotland, and local variants e.g. Moray skiff. Both bow and stern curve gracefully
inwards from amidships and taper to a pointed end.

too, the wings are long and comparatively slender, in others shorter and of greater breadth; in some there is an inflection resembling the bend of an elbow, in others there is a continuous swelling from the termination to the shoulder, where a sudden narrowing takes place immediately over the articulation. I had inferred somewhat too hurriedly, though perhaps naturally enough, that these wings or arms, with their strong sharp points and oar-like blades, had been at once paddles and spears,—instruments of motion and weapons of defence; and hence the mistake of connecting the creature with the Chelonia. I am informed by Agassiz, however, that they were weapons of defence only, which, like the occipital spines of the river bull-head, were erected in moments of danger or alarm, and at other times lay close by the creature's side; and that the sole instrument of motion was the tail, which, when covered by its coat of scales, was proportionally of a somewhat larger size than the tail shown in the print, which, as in the specimens from which it was taken, exhibits but the obscure and uncertain lineaments of the skeleton. The river bullhead, when attacked by an enemy, or immediately as it feels the hook in its jaws, erects its two spines at nearly right angles with the plates of the head, as if to render itself as difficult of being swallowed as possible. The attitude is one of danger and alarm; and it is a curious fact, to which I shall afterwards have occasion to advert, that in this attitude nine-tenths of the *Ptericthyes* of the Lower Old Red Sandstone are to be found   We read in the stone a singularly preserved story of the strong instinctive love of life, and

river bull-head: bullhead or miller's thumb, a freshwater fish armed with spines for defence; *Ptericthyes*: emended in 3rd edition to *Pterichthyes*, *ichthyes* being the Classical Greek plural form of *ichthys*, 'fish'.

of the mingled fear and anger implanted for its preservation,—" The champions in distorted postures
threat ;"—it presents us, too, with a wonderful record
of violent death, falling at once, not on a few individuals, but on whole tribes.

Next to the *Ptericthys* of the Lower Old Red I
shall place its cotemporary the *Coccosteus* of Agassiz,
a fish which, in some respects, must have closely resembled it. Both were covered with an armour of
thickly tuberculated bony plates, and both furnished
with a vertebrated tail. The plates of the one, when
found lying detached in the rock, can scarcely be distinguished from those of the other: there are the same
marks as in the plates of the tortoise, of accessions
of growth at the edges ; the same porous bony structure within, the same kind of tubercules without.
The forms of the creatures themselves, however, were
essentially different. I have compared the figure of
the *Ptericthys,* as shown in some of my better specimens, to that of a man with the head cut off by the
shoulders, one of the legs also wanting, and the arms
spread to the full. The figure of the *Coccosteus* I
would compare to a boy's kite (see Plate III. fig 1).
There is a rounded head, a triangular body, a long
tail attached to the apex of the triangle, and no arms.
The manner in which the plates are arranged on the
head is peculiarly beautiful ; but I am afraid I cannot adequately describe them. A ring of plates, like
the ring-stones of an arch, runs along what may be
called the hoop of the kite; the form of the key-
stone-plate is perfect; the shapes of the others are
elegantly varied, as if for ornament; and what would

'The champions in distorted postures threat': Olympic athletes described in odes
by the ancient Greek poet Pindar, imagined frozen into sculptural forms in Alexander
Pope's poem *The Temple of Fame* (1710 or 1711); violent death ... on whole tribes: as
in the Arabian story of the 'city of statues' (p. 12).

be otherwise the opening of the arch, is filled up with one large plate, of an outline singularly elegant. A single plate, still larger than any of the others, covers the greater part of the creature's triangular body, to the shape of which it nearly conforms. It rises saddle-wise towards the centre: on the ridge there is a longitudinal groove ending in a perforation, a little over the apex (Plate III. fig 2); two small lateral plates on either side fill up the base of the angle ; and the long vertebrated tail terminates the figure.

Does the reader possess a copy of Lyell's lately published elementary work, edition 1838 ? If so, let him first turn up the description of the upper Silurian rocks, from Murchison, which occurs in page 459, and mark the form of the trilobite *Asaphus caudatus,* a fossil of what is termed the Wenlock formation. (See *Sil. Sys.* Plate VII.) The upper part, or head, forms a crescent ; the body rises out of the concave with a sweep somewhat resembling that of a Gothic arch ; the outline of the whole approximates to that of an egg, the smaller end terminating in a sharp point. Let him remark, further, that this creature was a *crustaceous* animal, of the crab or lobster family, and then turn up the brief description of the Old Red Sandstone in the same volume, page 454, and mark the form of the *Cephalaspis,* or buckler-head,—a *fish* of a formation over that in which the remains of the trilobite most abound. He will find that the fish and crustacean are wonderfully alike. The fish is more elongated, but both possess the crescent-shaped head, and both the angular and

---

elementary work: *Elements of Geology*, Lyell 1838 (see p. 19); **Sil. Sys.**: *The Silurian System*, Murchison 1839.

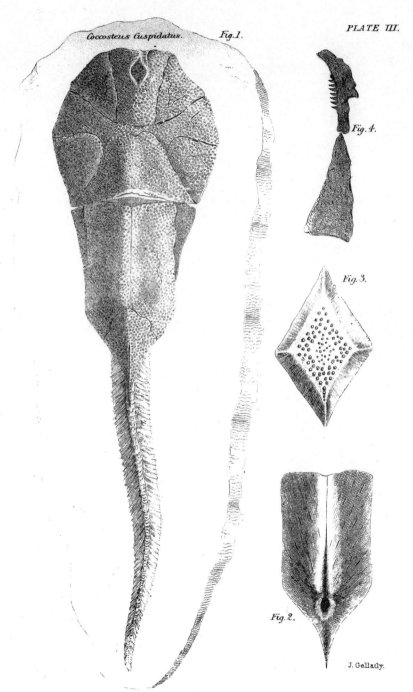

Coccosteus Cuspidatus.   *Fig. 1.*   PLATE III.

Fig. 4.

Fig. 3.

Fig. 2.

J. Gellatly.

Published by John Johnstone, Hunter Square, Edinburgh.

apparently jointed body.* They illustrate admirably how two distinct orders may meet. They exhibit the points, if I may so speak, at which the plated fish is linked to the shelled crustacean. Now, the *Coccosteus* is a stage further on; it is more unequivocally a fish. It is a *Cephalaspis* with a scale-covered tail attached to the angular body, and the horns of the crescent-shaped head cut off.

Some of the specimens which exhibit this creature are exceedingly curious. In one, a coprolite still rests in the abdomen; and a common botanist's microscope shows it thickly speckled over with minute scales, the indigestible exuviæ of fish on which the animal preyed. In the abdomen of another we find a few minute pebbles,—just as pebbles are occasionally found in the stomach of the cod,—which had been swallowed by the creature, attached to its food. Is there nothing wonderful in the fact that men should be learning at this time of day how the fishes of the Old Red Sandstone lived, and that there were some of them rapacious enough not to be over nice in their eating?

* Really jointed in the case of the trilobite; only apparently so in that of the *Cephalaspis*. The body of the trilobite, like that of the lobster, was barred by transverse oblong overlapping plates, and between every two plates there was a joint; the body of the *Cephalaspis*, in like manner, was barred by transverse oblong overlapping scales, between which there existed no such joints. It is interesting to observe how nature, in thus bringing two such different classes as fishes and crustacea together, gives to the higher animal a sort of pictorial resemblance to the lower, in parts where the construction could not be identical without interfering with the grand distinctions of the classes.

---

common botanist's microscope: essentially a simple lens or lenses and an adjustable mirror for light, mounted on a brass arm on a small wooden base which doubled as a pocket carrying-box; Miller had one with him when he found the fossil fishes at Cromarty (Miller 1854a, p. 452; Morrison-Low and Nuttall 2003); **abdomen**: as seen in the ichthyosaurs in Buckland 1836, vol. I, pp. 201–2. Buckland (p. 56) was renowned for his ability to reconstruct some extinct predators' feeding habits from their fossilised abdominal contents and droppings; **nice**: fussy.

The under part of the creature is still very imperfectly known: it had its central lozenge-shaped plate, like that on the under side of the *Pterichthys*, but of greater elegance (see Plate III. fig. 3), round which the other plates were ranged. " What an appropriate ornament, if set in gold," said Dr Buckland, on seeing a very beautiful specimen of this central lozenge in the interesting collection of Professor Traill of Edinburgh,—" what an appropriate ornament for a lady-geologist !" There are two marked peculiarities in the jaws of the *Coccosteus*, as shown in most of the specimens illustrative of the lower part of the creature, which I have yet seen. The teeth, instead of being fixed in sockets, like those of quadrupeds and reptiles, or merely placed on the bone, like those of fish of the common varieties, seem to have been cut out of the solid, like the teeth of a saw or the teeth in the mandibles of the beetle, or in the nippers of the lobster (Plate III. fig. 4) ; and there appears to have been something strangely anomalous in the position of the jaws,—something too anomalous, perhaps, to be regarded as proven by the evidence of the specimens yet found, but which may be mentioned with the view of directing attention to it. " Do not be deterred," said Agassiz, in the course of one of the interviews in which he obligingly indulged the writer of these chapters, who had mentioned to him that one of his opinions, just confirmed by the naturalist, had seemed so extraordinary that he had been almost afraid to communicate it,—" Do not be deterred, if you have ox amined minutely, by any dread of being deemed

---

'**What an appropriate ornament …**': presumably at Traill's breakfast party in Edinburgh, October 1840, Appendix 2; **Buckland**: William Buckland (1784–1856), Reader (effectively, professor) in Geology and in Mineralogy at the University of Oxford; a noted researcher on bone caves and the reconstruction of ancient vertebrates, and one of the most important British geologists of the day (Rupke 1983; Haile 2014); **Traill**: Thomas S. Traill (1781–1862), Professor of Medical Jurisprudence at the University of Edinburgh, curator of the Library and Museum of the Royal Society of Edinburgh, and collector of Old Red Sandstone fish: Andrews 1982a; Waterston 1997; White 2004;

extravagant. The possibilities of existence run so deeply into the extravagant, that there is scarcely any conception too extraordinary for nature to realize." In all the more complete specimens which I have yet seen, *the position of the jaws is vertical, not horizontal ;* and yet the creature, as shown by the tail, belonged unquestionably to the vertebrata. Now, though the mouths of the crustaceous animals, such as the crab and lobster, open vertically, and a similar arrangement obtains among the insect tribes, it has been remarked by naturalists as an invariable condition of that higher order of animals distinguished by vertebrated columns, that their mouths open horizontally. What I would remark as very extraordinary in the *Coccosteus,*—not, however, in the way of directly asserting the fact, but merely by way of soliciting inquiry regarding it,—is, that it seems to unite to a vertebral column a vertical mouth, thus forming a connecting link between two orders of existences, by conjoining what is at once their most characteristic and most dissimilar traits.

I am acquainted with four species of *Coccosteus,*— *C. latus, C. cuspidatus, C. oblongus,* and a variety not yet named ; and many more species may yet be discovered. Of all the existences of the formation, this curious fish seems to have been one of the most abundant. In a few square yards of rock I have laid open portions of the remains of a dozen different individuals belonging to two of the four species, the *C. latus* and *C. cuspidatus,* in the course of a single evening. None of the other kinds have yet been found at Cromarty. These two differed from each

---

proven (p. 56): Miller was right to be wary. Further specimens and research would rule out the 'vertical' mouth: Andrews 1982a, p. 23; **interviews** (p. 56): conversations.

vertical mouth: i.e. with the jaws working from side to side as in crustaceans, not up and down as in normal vertebrates; **a few square yards of rock**: presumably his Cromarty site.

other in the proportions which their general bulk bore to their length,—slightly too in the arrangement of their occipital plates. The *Coccosteus latus*, as the name implies, must have been by much a massier fish than the other; and we find the arch-like form of the plates which covered its head more complete: the plate representing the key-stone rests on the saddle-shaped plate in the centre, and the plates representing the spring-stones of the arch exhibit a broader base. The accompanying print (Plate III.) represents the *Coccosteus cuspidatus*. The average length of the creature, including the tail, as shown in most of the Cromarty specimens, somewhat exceeded a foot. A few detached plates from Orkney, in the collection of Dr Traill, must have belonged to an individual of fully twice that length.

CHAPTER IV.

The Elfin-fish of Gawin Douglas.—The Fish of the Old Red Sandstone scarcely less curious.—Place which they occupied indicated in the present Creation by a mere Gap.—Fish divided into two great Series,—the Osseous and Cartilaginous.—Their distinctive Peculiarities.—Geological Illustration of Dr Johnson's shrewd Objection to the Theory of Soame Jenyns.—Proofs of the intermediate Character of the Ichthyolites of the Old Red Sandstone.—Appearances which first led the Writer to deem it intermediate.—Confirmation by Agassiz.—The *Osteolepis.*—Order to which this Ichthyolite belonged.—Description.—*Dipterus.*—*Diplopterus.*—*Cheirolepis.*—*Glyptolepis.*

HAS the reader ever heard of the " griesly fisch" and the " laithlie flood," described by the minstrel Bishop of Dunkeld, " who gave rude Scotland Virgil's page ?" Both fish and flood are the extravagancies of a poet's dream. The flood came rolling through a wilderness of bogs and quagmires, under banks " dark as rocks the whilk the sey upcast." A skeleton forest stretched around, doddered and leafless ; and through the " unblomit" and " barrant" trees

" The quhissling wind blew mony bitter blast;"

the whitened branches " clashed and clattered ;" the

---

griesly fisch: frightful, ghastly fish (mediaeval Scots); **laithlie**: loathly, horrible; **Bishop of Dunkeld**: Gavin Douglas (*c.*1475–1522), third son of the 5th Earl of Angus. A great makar (Scots: poet). First translator of Virgil's *Aeneid* into Scots. The dream sequence quoted opens Douglas's courtly allegory *The Pallice of Honour*, **'who gave rude Scotland Virgil's page'**: Walter Scott, *Marmion*, canto 6. The other quotations on pp. 59–60 are all from the first four stanzas of *Pallice*, freely adapted: **'the whilk the sey upcast'**, which the sea cast up; **'unblomit'**, without blossom; **'barrant'**, barren; **'The quhissling wind blew mony bitter blast'**, the whistling wind blew many a bitter blast.

" vile water rinnand o'erheid," and " routing as thonder," made " hideous trubil;" and to augment the up-
roar, the " griesly fisch," like the fish of eastern story,
raised their heads amid the foam, and shrieked and
yelled as they passed. " The grim monsters fordeafit
the heiring with their schouts ;"—they were both fish
and elves, and strangely noisy in the latter capacity ;
and the longer the poet listened, the more frightened
he became. The description concludes, like a terri-
fic dream, with his wanderings through the labyrinths
of the dead forest, where all was dry and sapless
above, and mud and marsh below, and with his ex-
clamations of grief and terror at finding himself hope-
lessly lost in a scene of prodigies and evil spirits.
And such was one of the wilder fancies in which
a youthful Scottish poet of the days of Flodden indulged, ere taste had arisen to restrain and regulate
invention.

Shall I venture to say, that the ichthyolites of the
Old Red Sandstone have sometimes reminded me of
the " fisch of the laithlie flood ?" They were hardly
less curious. We find them surrounded, like these,
by a wilderness of dead vegetation and of rocks upcast from the sea ; and there are the footprints of
storm and tempest around and under them. True,
they must have been less noisy. Like the " griesly
fisch," however, they exhibit a strange union of opposite natures. One of their families—that of the
Cephalaspis—seems almost to constitute a connecting
link, says Agassiz, between fishes and crustaceons.
They had also their families of sauroid or reptile fishes
—and their still more numerous families that unite

'vile water rinnand o'erheid': foul water running overhead; 'routing as thonder':
roaring like thunder (Douglas: 'as thonder routit'); 'hideous trubil': hideous trouble;
Agassiz: either from *Poissons fossiles* or the translated quotations from it in *The Silurian
System*: Murchison 1839, part II, esp. pp. 595–96; **eastern story**: probably that mentioned on p. 101; 'fordeafit the heiring with their schouts': deafened the hearing with
their shouts (Douglas: 'thair zelpis wilde my heiring all fordeifit', their wild yelps completely deafened my hearing); **Flodden**: the disastrous defeat of the Scots by an
English army in 1513; **crustaceons**: emended in 3rd ed. to 'crustaceans'.

the cartilaginous fishes to the osseous. And to these last the explorer of the Lower Old Red Sandstone finds himself mainly restricted. The links of the system are all connecting links, separated by untold ages from that which they connect,—so that in searching for their representatives amid the existences of the present time, we find but the gaps which they should have occupied. And it is essentially necessary from this circumstance, in acquainting one's self with their peculiarities, to examine, if I may so express myself, the sides of these gaps,—the existing links at both ends to which the broken links should have pieced,—in short, all those more striking peculiarities of the existing disparted families which we find united in the intermediate families that no longer exist. Without some such preparation, the inquirer would inevitably share the fate of the poetical dreamer of Dunkeld, by losing his way in a labyrinth. In passing, therefore, with this object from the extinct to the recent, I venture to solicit, for a few paragraphs, the attention of the reader.

Fishes, the fourth great class in point of rank in the animal kingdom, and, in extent of territory, decidedly the first, are divided, as they exist in the present creation, into two distinct series,—the osseous and the cartilaginous. The osseous embraces that vast assemblage which naturalists describe as " fishes properly so called," and whose skeletons, like those of mammalia, birds, and reptiles, are composed chiefly of a calcareous earth, pervading an organic base. Hence the durability of their remains. In the cartilaginous series, on the contrary, the skeleton contains

---

disparted: separated from each other; **fourth great class**: in the chain of being, mammals were seen as the highest class of animal, followed by birds, reptiles (including amphibians), and fishes, before the rest; compare p. 62; **calcareous earth … organic base**: in modern terms, a calcium salt in a protein matrix.

scarce any of this earth; it is a framework of indurated animal matter, elastic, semi-transparent, yielding easily to the knife, and, like all mere animal substances, inevitably subject to decay. I have seen the
huge cartilaginous skeleton of a shark lost in a mass
of putrefaction in less than a fortnight. I have
found the minutest bones of the osseous ichthyolites
of the Lias entire after the lapse of unnumbered centuries.

The two series do not seem to precede or follow one
another in any such natural sequence as that in which
the great classes of the animal kingdom are arranged.
The mammifer takes precedence of the bird, the bird
of the reptile, the reptile of the fish; there is progression in the scale; the arrangement of the classes
is consecutive, not parallel.   But in this great division
there is no such progression; the osseous fish takes
no precedence of the cartilaginous fish, or the cartilaginous, as a series, of the osseous.   The arrangement
is parallel, not consecutive; but the parallelism, if I
may so express myself, seems to be that of a longer
with a shorter line;—the cartilaginous fishes, though
much less numerous in their orders and families than
the other, stretch farther along the scale in opposite
directions, at once rising higher and sinking lower
than the osseous fishes.   The cartilaginous order of the
sturgeons, a roe-depositing tribe, devoid alike of affection for their young, or of those attachments which
give the wild beasts of the forest partners in their
dens, may be regarded as fully abreast of by much
the greater part of the osseous fishes in both their
instincts and their organization.   The family of the

---

animal matter: organic material, in the biochemical sense; reptile: including amphibians (the modern 'herptile' is equivalent); parallelism: see pp. A69–70.

sharks, on the other hand, and some of the rays, rise higher, as if to connect the class of fish with the class immediately above it,—that of reptiles. Many of them are viviparous, like the mammalia,—attached, it is said, to their young, and fully equal to even birds in the strength of their connubial attachments. The male, in some instances, has been known to pine away and die when deprived of his female companion.* But then, on the other hand, the cartilaginous fishes in some of their tribes sink as low beneath the osseous as they rise above them in others. The suckers, for instance, a cartilaginous family, are the most imperfect of all vertebrated animals; some of them want even the sense of sight; they seem mere worms furnished with fins and gills, and were so classed by Linnæus; but though now ascertained to be in reality fishes, they must be regarded as the lowest link in the scale,—as connecting the class with the class *Vermes*, just as the superior cartilaginous

* Some of the osseous fishes are also viviparous,—the " viviparous blenny," for instance. The evidence from which the supposed affection of the higher fishes for their offspring has been inferred, is, I am afraid, of a somewhat equivocal character. The love of the sow for her litter hovers at times between that of the parent and that of the epicure; nor have we proof enough, in the present state of ichthyological knowledge, to conclude to which side the parental love of the fish inclines. The connubial affections of some of the higher families seem better established. Of a pair of gigantic rays *(Cephaloptera giorna)* taken in the Mediterranean, and described by Risso, the female was captured by some fishermen; and the male continued constantly about the boat, as if bewailing the fate of his companion, and was then found floating dead. (See Wilson's article ICHTHYOLOGY, *Encyc. Brit.* seventh edit.)

---

suckers: here, hagfishes and lampreys; **Linnaeus**: the formally Latinised name of Carl von Linné (1707–78), the great eighteenth-century Swedish naturalist who codified modern biological classification, using in particular the two-part 'binomen' combining genus and species names, e.g. *Pterichthys milleri*; **Vermes**: old zoological name for a group containing the worms; **Encyc. Brit.**: *Encyclopaedia Britannica*.

fishes may be regarded as connecting it with the class *Reptilia*.

Between the osseous and the cartilaginous fishes there exist some very striking dissimilarities. The skull of the osseous fish is divided into a greater number of distinct bones, and possesses more moveable parts, than the skulls of mammiferous animals: the skull of the cartilaginous fish, on the contrary, consists of but a single piece, without joint or suture. There is another marked distinction. The bony fish, if it approaches in form to that general type which we recognise amid all the varieties of the class as proper to fishes, and to which in all their families nature is continually inclining, will be found to have a tail branching out, as in the perch and herring, from the bone in which the vertebral column terminates ; whereas the cartilaginous fish, if it also approach the general type, will be found to have a tail formed, as in the sturgeon and dog-fish, on both sides of the lower portion of the spine, but developed much more largely on the under than on the upper side. In some instances it is wanting on the upper side altogether. It may be as impossible to assign reasons for such relations as for those which exist between the digestive organs and the hoofs of the ruminant animals ; but it is of importance that they should be noted.* It may be remarked, further, that

---

* Dr Buckland, in his *Bridgewater Treatise*, assigns satisfactory reasons for this construction of tail in sharks and sturgeons. Of the fishes of these two orders, he states, " the former perform the office of scavengers, to clear the water of impurities, and have no teeth, but feed, by means of a soft

---

spine: here the vertebral column or backbone; in this context **lower** means 'posterior' or 'terminal'; **ruminant**: odd-toed ungulates such as horses have simple fore-stomachs, in contrast to those even-toed ungulates such as cattle which have complex multi-chambered stomachs and chew the cud; ***Bridgewater Treatise***: Buckland's *Geology and Mineralogy* (1836); see pp. A42, A57.

the great bulk of fishes whose skeletons consist of cartilage, have yet an ability of secreting the calcareous earth which composes bone, and that they are furnished with bony coverings, either partial or entire. Their bones lie outside. The thorn-back derives its name from the multitudinous hooks and spikes of bone that bristle over its body; the head, back, and operculum of the sturgeon are covered with bony plates; the thorns and prickles of the shark are composed of the same material. The framework within is a framework of mere animal matter; but it was no lack of the osseous ingredient that led to the arrangement,— an arrangement which we can alone refer to the will of that all-potent Creator, who can transpose his materials at pleasure, without interfering with the perfection of his work. It is a curious enough circumstance, that some of the osseous fishes, as if entirely to reverse the condition of the cartilaginous ones, are partially covered with plates of cartilage. They are bone within and cartilage without, just as the others are bone without and cartilage within.

But how apply all this to the Geology of the Old Red Sandstone? Very directly. The ichthyolites of

leather-like mouth, capable of protrusion and contraction, on putrid vegetables and animal substances at the bottom, and hence they have constantly to keep their bodies in an inclined position. The sharks employ their tail in another peculiar manner, to turn their body, in order to bring their mouth, which is placed downwards beneath the head, into contact with their prey. We find an important provision in every animal to give a position of ease and activity to the head during the operation of feeding." (*Bridgewater Treatise*, p. 279, vol. i. first edit.)

E

---

thorn-back: thornback ray, one of the fishmonger's 'skates'; **animal matter**: i.e. unmineralised cartilage.

this ancient formation hold, as has been said, an intermediate place, unoccupied among present existences, between the two series, and in some respects resemble the osseous, and in some the cartilaginous tribes. The fact reminds one of Dr Johnson's shrewd objection to the theory embraced by Soame Jenyns in his *Free Inquiry*, and which was the theory also of Pope and Bolingbroke. The metaphysician held with the poet and his friend, that there exists a vast and finely-graduated chain of being from Infinity to nonentity—from God to nothing; and that to strike out a single link would be to mar the perfection of the whole.* The moralist demonstrated, on the contrary, that this chain, in the very nature of things, must be incomplete at both ends,—that between that which does, and that which does not exist, there must be an infinite difference,—that the chain, therefore, cannot lay hold on *nothing*. He showed, further, that between the greatest of finite existences and

* " See, through this air, this ocean, and this earth,
    All matter quick and bursting into birth;
    Above how high progressive life may go !
    Around how wide ! how deep extend below!
    Vast chain of being ! which from God began—
    Nature's ethereal, human angel, man,
    Beast, bird, fish, insect, what no eye can see,
    No glass can reach; from infinite to Thee,—
    From Thee to nothing. On superior powers
    Were we to press, inferior might on ours;
    Or in the full creation leave a void,
    Where, one step broken, the great scale's destroyed :
    From Nature's chain whatever link you strike,
    Tenth, or ten thousandth, breaks the chain alike."
                                    (*Essay on Man.*)

Johnson: Samuel Johnson (1709–84), English man of letters, lexicographer, critic and **moralist**; **Soame Jenyns**: (1704–87), English politician, philosopher ('**metaphysician**') and author of *A Free Inquiry into the Nature and Origin of Evil* (1757); **Pope**: Alexander Pope (1688–1744), English poet; Miller quotes from *Essay on Man* (published 1733–34), Epistle I; **Bolingbroke**: Henry St John, 1st Viscount Bolingbroke (1678–1751), English statesman (and **friend** of Pope) who discussed the chain of being in his *Philosophical Fragments*, written between 1726 and 1734.

the adorable Infinite there must exist another illimit-
able void,—that the boundless and the bounded are
as widely separated in their natures and qualities, as
the existent and the non-existent,—that the chain, in
short, cannot lay hold on Deity. He asserted, however, that not only is it thus incomplete at both ends,
but that we must regard it as well nigh as incomplete
in many of its intermediate links, as at its terminal
ones,—that it is already a broken chain, seeing that
between its various classes of existence, myriads of
intermediate existences might be introduced, by graduating more minutely what must necessarily be
capable of infinite gradation,—and that, to base an
infidel theory on the supposed completeness of what
is demonstrably incomplete, and on the impossibility
of a gap existing in what is already filled with gaps,
is just to base one absurdity on another.*· Now, we

* The following are the well-stated reasonings of Dr Johnson, a writer who never did injustice to an argument for want
of words to express it in.

" The scale of existence from infinity to nothing cannot
possibly have being. The highest being not infinite must be
at an infinite distance from infinity. Cheyne, who, with the
desire inherent in mathematicians to reduce every thing to
mathematical images, considers all existence as a *cone*, allows
that the basis is at an infinite distance from the body, and in
this distance between finite and infinite there will be room for
ever for an infinite series of indefinable existence.

" Between the lowest positive existence and nothing, whenever we suppose positive existence to cease, is another chasm
infinitely deep, where there is room again for endless orders
of subordinate nature, continued for ever and ever, and yet
infinitely superior to non-existence.

" To these meditations humanity is unequal. But yet we
may ask, not of our Maker, but of each other, since on the

find the Geology of what may be termed the second
age of vertebrated existence (for the Lower Old Red
Sandstone was such) coming curiously in to confirm
the reasonings of Johnson. It shows us the greater
part of the fish of an entire creation thus insinuated
between two of the links of our own.

It is now several years since I was first led to sus-
pect that the condition of the ichthyolites of the Old
Red Sandstone was intermediate. I have alluded to
the comparative indestructibility of the osseous skeleton, and the extreme liability to decay characteristic
of the cartilaginous one. Of a skeleton in part osseous and in part cartilaginous, we must, of course, expect, when it occurs in a fossil state, to find the inde-
structible portions only. And when, in every instance, we find the fossil skeletons of a formation
complete in some of their parts, and incomplete in
others,—the entire portions invariably agreeing, and

one side creation, whenever it stops, must stop infinitely below infinity, and on the other infinitely above nothing, what
necessity there is that it should proceed so far either way,—
that being so high or so low should ever have existed. We
may ask, but I believe no created wisdom can give an adequate answer.

" Nor is this all. In the scale, wherever it begins or ends,
are infinite vacuities. At whatever distance we suppose the
next order of beings to be above man, there is room for an intermediate order of beings between them; and if for one order, then for infinite orders; since every thing that admits of
more or less, and consequently all the parts of that which admits them, may be infinitely divided, so that, as far as we can
judge, there may be room in the vacuity between any two
steps of the scale, or between any two points of the cone of
being, for infinite exertion of infinite power." (Remain of "A
Free Enquiry.")

---

entire creation: i.e. separate geological period; but see pp. A61–63; **Johnson** (also
quotations in footnote): from his 'Review of *A Free Inquiry into the Nature and Origin
of Evil* '.

the wanting portions invariably agreeing also,—it seems but natural to conclude, that an original difference must have obtained, and that the existing parts, which we can at once recognize as bone, must have been united to parts now wanting, which were composed of cartilage. The naturalist never doubts that the shark's teeth which he finds detached on the shore, or buried in some ancient formation, were united originally to cartilaginous jaws. Now, in breaking open all the ichthyolites of the Lower Old Red Sandstone, with the exception of those of the two families already described, we find that some of the parts are invariably wanting, however excellent the state of preservation maintained by the rest. I have seen every scale preserved and in its place,—one set of both the larger and smaller bones occupying their original position,—jaws thickly set with teeth still undetached from the head,—the massy bones of the scull still unseparated, the larger shoulder-bone, on which the operculum rests, lying in its proper bed, —the operculum itself entire,—and all the external rays which support the fins, though frequently fine as hairs, spreading out distinct as the fibres in the wing of the dragon-fly, or the woody nerves in an oak leaf. In no case, however, have I succeeded in finding a single joint of the vertebral column, or the trace of a single internal ray. No part of the internal skeleton survives, nor does its disappearance seem to have had any connection with the greater mass of putrescent matter which must have surrounded it, seeing that the external rays of the fins show quite as

---

scull: skull; **single joint of the vertebral column**: individual vertebra; **internal ray**: one of the vertical 'processes' and lateral ribs radiating from the backbone, and the similar bones in the bases of certain fins.

entire when turned over upon the body, as sometimes occurs, as when spread out from it in profile.
Besides, in the ichthyolites of the chalk, no parts of
the skeleton are better preserved than the internal
parts,—the vertebral joints and the internal rays.
The reader must have observed, in the cases of a
museum of Natural History, preparations of fish of
two several kinds,—preparations of the skeleton, in
which only the osseous parts are exhibited, and preparations of the external form, in which the whole
body is shown in profile, with the fins spread to the
full, and at least half the bones of the head covered
by the skin, but in which the vertebral column and
internal rays are wanting.   Now, in the fossils of the
chalk, with those of the other later formations, down
to the New Red Sandstone, we find that the skeleton
style of preparation obtains ; whereas in at least
three-fourths of the ichthyolites of the Lower Old
Red, we find only what we may term the external
style.   I had marked, besides, another circumstance
in the ichthyolites, which seemed, like a nice point of
circumstantial evidence, to give testimony in the same
line.   The tails of all the ichthyolites, whose vertebral columns and internal rays are wanting, are unequally lobed, like those of the dog-fish and sturgeon
(both cartilaginous fishes), and the body runs on to
nearly the termination of the surrounding rays.   The
one-sided condition of tail exists, says Cuvier, in no
recent osseous fish known to naturalists except in
the bony pike,—a sauroid fish of the warmer rivers
of America.   With deference, however, to so high

---

preparation: mounting for preservation and display, one technique being to make a
dried skin in three dimensions, often with one side missing and the interior completely
empty. Today, painted casts are used; **bony pike**: not the British pike, but the unrelated
gar or garpike found in North America.

an authority, it is questionable whether the tail of the bony pike should not rather be described as a tail set on somewhat awry, than as a one-sided tail.

All these peculiarities I could but note as they turned up before me, and express, in pointing them out to a few friends, a sort of vague, because hopeless, desire that good fortune might throw me in the way of the one man of all the world best qualified to explain the principle on which they occurred, and to decide whether fishes may be at once bony and cartilaginous. But that meeting was a contingency rather to be wished than hoped for,—a circumstance within the bounds of the possible, but beyond those of the probable. Could the working man of the north of Scotland have so much as dreamed that he was yet to enjoy an opportunity of comparing his observations with those of the naturalist of Neufchatel, and of having his inferences tested and confirmed ?

The opportunity did occur. The working man did meet with Agassiz ; and many a query had he to put to him ; and never surely was inquirer more courteously entreated, or his doubts more satisfactorily resolved. The reply to almost my first question solved the enigma of nearly ten years' standing. And finely characteristic was that reply of the frankness and candour of a great mind, that can afford to make it no secret, that in its onward advances on knowledge it may know to-day what it did not know yesterday, and that it is content to " gain by degrees upon the darkness." " Had you asked me the question a fortnight ago," said Agassiz, " I could not have replied to it. Since then, however, I have examined an ich-

---

meet: in October 1840; 'gain by degrees upon the darkness': Miller recalls a much-quoted phrase first used in Samuel Johnson's *Lives of Eminent Persons* to emphasise that, even for geniuses, scientific insights are usually approached in small steps rather than in sudden leaps: 'even the mind of [Isaac] Newton gains ground gradually upon darkness'. 'By degrees' appears a few lines later in the same essay; system: probably from Agassiz 1834 (or a version of this such as Buckland 1836, vol. I, pp. 273–74), as his better-known essay on classification was not published before 1843 (Agassiz 1833–43, vol. I, pp. 165–72; Brown 1890, p. xxv; Jeannet 1928, p. 122).

thyolite of the Old Red Sandstone in which the vertebral joints are fortunately impressed on the stone,
though the joints themselves have disappeared, and
which, exactly resembling the vertebræ of the shark,
must have been cartilaginous." In a subsequent conversation the writer was gratified by finding most
of his other facts and inferences authenticated and
confirmed by those of the naturalist. I shall attempt introducing to the reader the peculiarities, general and specific, of the ichthyolites to which these
facts and observations mainly referred, by describing
such of the families as are most abundant in the formation, and the points in which they either resemble
or differ from the existing fish of our seas.

Of these ancient families, the *Osteolepis*, or bony-
scale (see Plate IV. fig. 1), may be regarded as
illustrative of the general type. It was one of the
first discovered of the Caithness fishes, and received
its name, in the days of Cuvier, from the osseous
character of its scales, ere it was ascertained that it
had had numerous cotemporaries, and that to all and
each of these the same description applied. The
scales of the fishes of the Lower Old Red Sandstone,
like the plates and detached prickles of the purely
cartilaginous fishes, were composed of a bony, not of
a horny substance, and were all coated externally
with enamel. The circumstance is one of interest.

Agassiz, in his system of classification, has divided
fishes into four orders, according to the form of their
scales ; and his principle of division, though apparently arbitrary and trivial, is yet found to separate
the class into great natural families, distinguished

---

received its name: during work by two Paris-based palaeontologists, Achille Valenciennes and Joseph Barclay Pentland, as published in Sedgwick and Murchison 1829b,
pp. 143–44 (which, however, credits Valenciennes alone with the genus name); **horny**:
Miller is comparing the scales of *Osteolepis* to the flexible scales of many modern fishes,
principally the teleost fishes such as salmon and cod.

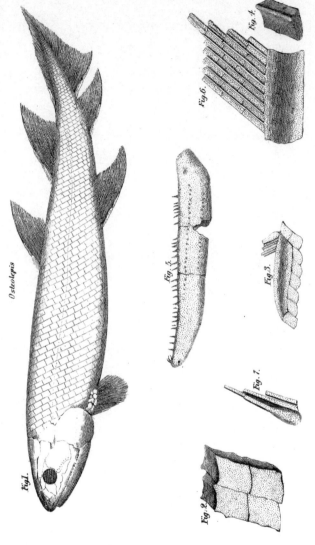

PLATE IV.

Osteolepis

Fig. 1.

Fig. 2.

Fig. 3.

Fig. 4.

Fig. 5.

Fig. 6.

Fig. 7.

J. Gellatly.

Published by John Johnstone, Hunter Square, Edinburgh.

from one another by other and very striking peculiarities. One kind of scale, for instance the Placoid or broad-plated scale, is found to characterize all the cartilaginous fishes of Cuvier except the sturgeon ;—it is the characteristic of an otherwise well-marked series, whose families are furnished with skeletons composed of mere animal matter, and whose gills open to the water by spiracles. The fish of another order are covered by ctenoid or comb-shaped scales, the posterior margin of each scale being toothed somewhat like the edge of a saw or comb; and the order, thus distinguished, is found wonderfully to agree with an order formed previously on another principle of classification, the Acanthopterygii, or thorny-finned order of Cuvier, excluding only the smooth-scaled families of this previously-formed division, and including, in addition to it, the flat fish. A third order, the Cycloidean, is marked by simple marginated scales, like those of the cod, haddock, whiting, herring, salmon, &c.; and this order is found to embrace chiefly the Malacopterygii, or soft-finned order of Cuvier,—an order to which all these well-known fish, with an immense multitude of others, belong. Thus the results of the principle of classification adopted by Agassiz wonderfully agree with the results of the less simple principles adopted by Cuvier and the other masters in this department of Natural History. Now, it is peculiar to yet a fourth order, the Ganoidean or shining-scaled order, that by much the greater number of the genera which it comprises exist only in the fossil state. At least five-sixths of the whole

were ascertained to be extinct several years ago, at a
time when the knowledge of fossil Ichthyology was
much more limited than at present : the proportions
are now found to be immensely greater on the side of
the dead.   And this order seems to have included all
the semi-osseous semi-cartilaginous ichthyolites of the
Lower Old Red Sandstone ; the enamelled scale is
the characteristic, according to Agassiz's principle of
classification, of the existences that filled the gap so
often alluded to as existing in the present creation.
All their scales glitter with enamel : they bore to this
order the relation that the cartilaginous fish bear to
the Placoidean order, the thorny-finned fish to the
Ctenoidean order, and the soft-finned fish to the
Cycloidean order.   It also included, with the semi-
cartilaginous, the sauroid fish,—those master-existences and tyrants of the earlier vertebrata ; and both
classes find their representatives among the comparatively few ganoid fishes of the present creation ; the
one in the sturgeon family, which of all existing families approaches nearest in other respects to the extinct semi-cartilaginous fishes ; the other in the sauroid genus *Lepidosteus*, to which the bony pike belongs.   The head, back, and sides of the sturgeon
are defended, as has been already remarked, by longitudinal rows of hard osseous bosses,—the bony pike
is armed with enamelled osseous scales, of a stony
hardness. It seems a somewhat curious circumstance,
that fishes so unlike each other in their internal
frame-work, should thus resemble one another in
their bony coverings, and in some slight degree in

---

several years ago: probably taken from Buckland 1836, vol. I, pp. 269–70.

their structure of tail. One of the characteristics of sauroid fishes is the extreme compactness and hardness of their skeleton.*

It requires skill such as that possessed by Agassiz to determine that the uncouth *Coccosteus*, or the equally uncouth *Ptericthys*, of the Old Red Sandstone, with their long articulated tails and tortoise-like plates, were *bona fide* fishes; but there is no possibility of mistaking the *Osteolepis*: it is obvious to the least-practised eye that it must have been a fish, and a handsome one. Even a cursory examination, however, shows very striking peculiarities, which are found, on further examination, to characterise not this family alone, but at least one-half the cotemporary families besides. We are accustomed to see vertebrated animals with the bone uncovered in one part only,—that part the teeth,—and with the rest of the skeleton wrapped up in flesh and skin. Among the reptiles we find a few exceptions; but a creature with a skull as naked as its teeth,—the bone being merely covered, as in these, by a hard shining ena-

* "The sauroid or lizard-like fishes," says Dr Buckland, "combine in the structure, both of the bones and some of the soft parts, characters which are common to the class of reptiles. The bones of the skull are united by closer sutures than those of common fishes. The vertebræ articulate with the spinous processes by sutures, like the vertebræ of saurians; the ribs also articulate with the extremities of the spinous process. The caudal vertebræ have distinct chevron bones; and the general condition of the skeleton is stronger and more solid than in other fishes; the air bladder also is bifid and cellular, approaching to the character of lungs; and in the throat there is a glottis, as in sirens and salamanders, and many saurians. (Note to *Bridgewater Treatise*, p. 274, first edit.)

articulate with: join with; **Buckland**: Buckland 1836, vol. I, pp. 273–74, edited somewhat, mistranscribing and garbling the original text's 'with the extremities of the spinous processes'. Buckland was referring to bony bars or protrusions ('processes') fixed to the body of the vertebra by rigid joints ('sutures').

mel,—and with toes also of bare enamelled bone, would be deemed an anomaly in creation. And yet such was the condition of the *Osteolepis*, and many of its cotemporaries. The enamelled teeth were placed in jaws which presented outside a surface as naked and as finely enamelled as their own. The entire head was covered with enamelled osseous plates, furnished inside like other bones, as shown by their cellular construction, with their nourishing blood-vessels, and perhaps their oil, and which rested apparently on the cartilaginous box, which must have enclosed the brain, and connected it with the vertebral column. I cannot better illustrate the peculiar condition of the fins of this ichthyolite than by the webbed foot of a water-fowl. The web or membrane in all the aquatic birds with which we are acquainted, not only connects, but also covers the toes. The web or membrane in the fins of existing fishes accomplishes a similar purpose; it both connects and covers the supporting bones or rays. Imagine, however, a webbed foot in which the toes—connected but not covered—present, as in skeletons, an upper and under surface of naked bone; and a very correct idea may be formed from such a foot, of the condition of fin which obtained among at least one-half the ichthyolites of the Lower Old Red Sandstone. The supporting bones or rays were connected laterally by the membrane; but on both sides they presented bony and finely-enamelled surfaces. In this singular class of fish, all was bone without, and all was cartilage within, and the bone in every instance, whether in the form of jaws or of plates, of

scales or of rays, presented an external surface of enamel.

The fins are quite a study. I have alluded to
the connecting membrane. In existing fish this
membrane is the principal agent in propelling the
creature : it strikes against the water, as the membrane of the bat's wing strikes against the air; and the
internal skeleton serves but to support and stiffen it
for this purpose. But in the fin of the *Osteolepis*, as
in those of many of its cotemporaries, we find the
condition reversed. The rays were so numerous, and
lay so thickly side by side, like feathers in the wing
of a bird, that they presented to the water a continuous surface of bone, and the membrane only served
to support and bind them together. In the fins of
existing fish we find a sort of bat-wing construction,
—in those of the *Osteolepis* a sort of bird-wing construction. The rays, to give flexibility to the organ
which they compose, were all jointed, as in the soft-
finned fish,—as in the herring, salmon, and cod, for
example ; and we find in all the fins the anterior ray
rising from the body in the form of an angular scale :
it is a strong bony scale in one of its joints, and a
bony ray in the rest. The characteristic is a curious
one.

It is again necessary, in pursuing our description,
to refer for illustration to the purely cartilaginous
fishes. In at least all the higher orders of these, furnished with moveable jaws, such as the sturgeon, the
ray, and the shark, the mouth is placed far below the
snout. The dog-fish and thorn-back are familiar instances. Further, the mouth in bony fishes is move-

able on both the upper and under side, like the beak of the parrot ; in the higher cartilaginous fishes it is moveable, as in quadrupeds, on the under side only. In all their orders, too, except in that of the sturgeon, the gills open to the water by detached spiracles or breathing holes ; but in the sturgeon, as in the osseous fishes, there is a continuous linear opening, shielded by an operculum or gill-cover. In the *Osteolepis* the mouth opened below the snout, but not so far below it as in the purely cartilaginous fishes,—not farther below it than in many of the osseous ones,—than in the genus Aspro for instance, or than in the genus Polynemus, or in even the haddock or cod. It was thickly furnished with slender and sharply-pointed teeth. I have hitherto been unable fully to determine whether, like the mouths of the osseous fishes, it was moveable on both sides, though, from the perfect form of what seems to be the intermaxillary bone, I cannot avoid thinking it was. The gills opened, as in the osseous fishes, in continuous lines, and were covered by large bony opercules, that on the enamelled side somewhat resemble round japanned shields.

But while the head of the *Osteolepis*, with its appendages, thus resembled, in some points, the heads of the bony fishes, the tail, like those of most of its cotemporaries, differed in no respect from the tails of cartilaginous ones, such as the sturgeon. The vertebral column seems to have run on to well nigh the extremity of the caudal fin, which we find developed chiefly on the under side. The tail was a onesided tail. Take into account with these peculiari-

upper and under side: i.e. both upper and lower jaws are flexibly mounted on the skull.

ties,—peculiarities such as the naked skull, jaws, and operculum, the naked and thickly-set rays, and the unequally-lobed condition of tail,—a body covered with scales that glitter like sheets of mica, and assume, according to their position, the parallelogrammical, rhomboidal, angular, or polygonal form,—a lateral line raised, not depressed,—a raised bar on the inner or bony side of the scales, which, like the doubled-up end of a tile, seems to have served the purpose of fastening them in their places,—a general clustering of alternate fins towards the tail,—and the *tout ensemble* must surely impart to the reader the idea of a very singular little fish. The ventral fins front the space which occurs between the two dorsals, and the anal fin the space which intervenes between the posterior dorsal fin and the tail. The length of the *Osteolepis*, in my larger specimens, somewhat exceeds a foot; in the smaller it falls short of six inches. There exist at least three species of this ichthyolite, distinguished chiefly, in two of the instances, by the smaller and larger size of their scales, compared with the bulk of their bodies, and by punctulated markings on the enamel in the case of the third. This last, however, is no specific difference, but common to the entire genus, and to several other genera besides. The names are *Osteolepis Macrolepidotus, O. Microlepidotus,* and *O. Arenaceus.*

Next to the *Osteolepis* we may place the *Dipterus,* or double-wing, of the Lower Old Red Sandstone, an ichthyolite first introduced to the knowledge of geologists by Mr Murchison, who, with his friend Mr Sedgwick, figured and described it in a masterly

---

three species: perhaps taken from Murchison 1839, part II, p. 601; **Sedgwick**: Adam Sedgwick (1785–1873), Professor of Geology at the University of Cambridge and leading member of the Geological Society of London, was a major stratigrapher of the older rocks, collaborating with Murchison on the Old Red Sandstone of Scotland and the equivalent strata of Devon (Sedgwick and Murchison 1829b; Rudwick 1985; J. Secord 1986, 2013).

paper on the older sedimentary formations of the
north of Scotland, which appeared in the *Transactions
of the Geological Society of London* for 1828.
The name, derived from its two dorsals, would suit
equally well, like that of the *Osteolepis*, many of its
more recently-discovered cotemporaries. From the
latter ichthyolite it differed chiefly in the position
of its fins, which were opposite, not alternate ; the
double dorsals exactly fronting the anal and ventral
fins (see Plate V. fig. 1). The *Diplopterus*, a nearly
resembling ichthyolite of the same formation, also
owes its name to the order and arrangement of its
fins, which, like those of the *Dipterus*, were placed
fronting each other, and in pairs. But the head, in
proportion to the body, was of greater size than in
either the *Dipterus* or *Osteolepis ;* and the mouth, as
indicated by the creature's length of jaw, must have
been of much greater width. In their more striking
characteristics, however, the three genera seem to have
nearly agreed. In all alike, scales of bone glisten with
enamel; their jaws, enamel without and bone within,
bristle thick with sharp-pointed teeth; closely-jointed
plates, burnished like ancient helmets, cover their
heads, and seem to have formed a kind of outer table
to skulls externally of bone and internally of cartilage;
their gill-covers consist each of a single piece, like
the gill-cover of the sturgeon ; their tails were formed
chiefly on the lower side of their bodies; and the
rays of their fins, enamelled like their plates and their
scales, stand up over the connecting membrane, like
the steel or brass in that peculiar armour of the middle
ages, whose multitudinous pieces of metal were fast-

paper: Sedgwick and Murchison 1829b, pp. 142–43, plate 16; ***Diplopterus***: the fin
positions are probably taken from the brief description in Agassiz 1833–43, vol. II, part I,
p. 113. The relevant part-work was actually published in 1833 or 1835 (Brown 1890, p. xxv;
Woodward and Sherborn 1890, p. 68; Jeannet 1928, p. 120). Murchison (1839, part I,
p. 601) published a translation. However, Miller's account does not merely stem from
this account, or Agassiz's earlier and shorter descriptions (Agassiz 1835).

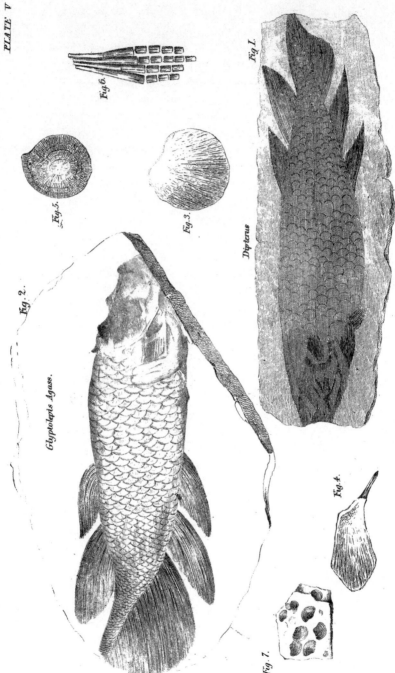

PLATE V

Fig.6.

Fig.1.

Dipterus

Fig.5.

Fig.3.

Fig.2.

Glyptolepis Agass.

Fig.4.

Fig.7.

J. Gellatly

Published by John Johnstone, Hunter Square, Edinburgh.

ened together on a groundwork of cloth or of leather. All their scales, plates, and rays, present a similar style of ornament. The shining and polished
enamel is mottled with thickly-set punctures, or rather punctulated markings; so that a scale or plate,
when viewed through a microscope, reminds one of
the cover of a saddle. Some of the ganoid scales of
Burdie House present surfaces similarly punctulated.

The *Glyptolepis*, or carved scale, may be regarded
as the representative of a family of the Lower Old
Red Sandstone, which, differing very materially from
the genera described, had yet many traits in common
with them, such as the bare bony skull, the bony
scales, the naked rays, and the unequally-sided condition of tail. The fins, which were of considerable
length in proportion to their breadth of base, and present in some of the specimens a pendulous-like appearance, cluster thick together towards the creature's
lower extremities, leaving the upper portion bare.
There are two dorsals placed as in the *Dipterus* and
Diplopterus,—the anterior directly opposite the ventral
fin, the posterior as directly opposite the anal. The
tail is long and spreading;—the rays, long and nu-
merously articulated, are comparatively stout at their
base, and slender as hairs where they terminate. The
shoulder-bones are of huge dimensions, the teeth extremely minute. But the most characteristic parts
of the creature are the scales. They are of great size
compared with the size of the animal. An individual not more than half a foot in length, the specimen figured (see Plate V. fig. 2), exhibits scales fully
three-eighth parts of an inch in diameter. In an-

F

other more broken specimen there are scales a full inch across, and yet the length of the ichthyolite to which they belonged seems not to have much exceeded a foot and a half. Each scale consists of a double plate, an inner and an outer. The structure of the inner is not peculiar to the family or the formation : it is formed of a number of minute concentric circles crossed by still minuter radiating lines,—the one described, and the other proceeding from a common centre. All scales that receive their accessions of growth equally at their edges, exhibit internally a corresponding character. The outer plate presents an appearance less common. It seems relieved into ridges that drop adown it like sculptured threads, some of them entire, some broken, some straight, some slightly waved ; and hence the name of the ichthyolite. The plates of the head were ornamented in a similar style, but their threads are so broken as to present the appearance of dotted lines, the dots all standing out in bold relief. My collection contains three varieties of this family; one of them disinterred from out the Cromarty beds about seven years ago, and the others only a little later, though, partly from the inadequacy of a written description, through which I was led to confound the *Osteolepis* with the *Diplopterus*, and to regard the *Glyptolepis* as the *Osteolepis*, I was not aware until lately that the discovery was really such ; and under the latter name I described the creature in the *Witness* newspaper several weeks ere it had received the name which it now bears. It was first introduced to the notice of Agassiz in autumn last by Lady Cum-

---

described ... proceeding: i.e. as circumference and radius from a common centre; **written description**: perhaps Sedgwick and Murchison 1829b, which published *Osteolepis* without illustrating it, or Agassiz's brief accounts of *Diplopterus* (see note for p. 80); **I was not aware**: Miller telescopes events of the previous seven years. In Cromarty, whatever start he had made with sorting his fossils, he would not have known all those genera, whether formally named or not, until contacts with other geologists from 1837 on; **in the *Witness***: [Miller] 1840f; **received the name**: when Agassiz visited Lady Gordon Cumming in October 1840, he recognised the new

ming of Altyre. The species, however, was a different one from any yet found at Cromarty.

The *Cheirolepis*, or scaly pectoral, forms the representative of yet another family of the Lower Old Red Sandstone, and one which any eye, however unpractised, could at once distinguish from the families just described. Professor Traill of the University of Edinburgh, a gentleman whose researches in Natural History have materially extended the boundaries of knowledge, and whose frankness in communicating information is only equalled by his facility in acquiring it, was the first discoverer of this family, one variety of which, the *Cheirolepis Traillii*, bears his name. The figured specimen (Plate VI. fig. 1) Agassiz has pronounced a new species, the discovery of the writer. In all the remains of this curious fish which I have hitherto seen, the union of the osseous with the cartilaginous, in the general framework of the creature, is strikingly apparent. The external skull, the great shoulder-bone, and the rays of the fins, are all unequivocally osseous ; the occipital and shoulder-bones in particular seem of great strength and massiveness, and are invariably preserved, however imperfect the specimen in other respects; whereas, even in specimens the most complete, and which exhibit every scale and every ray, however minute, and show unchanged the entire outline of the animal, not a fragment of the internal skeleton appears. The *Cheirolepis* seems to have varied from fourteen to four inches in length. When seen in profile, the under line, as in the figured variety, seems thickly covered with fins, and the upper line well nigh naked. The large pectorals almost

---

genus *Glyptolepis* and named it, but, did not himself publish the name in print till 1844 (Woodward and Sherborn 1890, p. 85; Andrews 1982a, pp. 28, 72 n. 123).

species … different: probably *Glyptolepis leptopterus*: Agassiz 1843b, p. 87; **scaly pectoral**: correctly 'scaly hand', see p. 89; **new species**: Agassiz later named it *Cheirolepis cummingiae*, from a Gordon Cumming specimen; the new name was added to Plate VI from the 3rd edition. Agassiz (1843b, p. 87) mentioned Cromarty as a source locality.

encroach on the ventral fins, and the ventrals on the
anal fin; whereas the back, for two-thirds the entire
length of the creature, presents a bare rectilinear
ridge, and the single dorsal, which rises but a little
way over the tail, immediately opposite the posterior
portion of the anal fin, is comparatively of small size.
The tail, which, in the general condition of being developed chiefly on the lower side, resembles the tails
of all the creature's cotemporaries, is elegantly lobed.
The scales, in proportion to the bulk of the body which
they cover, are not more than one-twentieth the size
of those of the *Osteolepis*. They are richly enamelled,
and range diagonally from the shoulder to the belly in
waving lines; and so fretted is each individual scale,
by longitudinal grooves and ridges, that on first
bringing it under the glass, it seems a little bunch of
glittering thorns, though, when more minutely examined, it is found to present somewhat the appearance of the outer side of the deep-sea cockle, with its
strongly-marked ribs and channels, the point in
which the posterior margin terminates representing the
hinge. The bones of the head, enamelled like the
scales, are carved into jagged inequalities, somewhat
resembling those on the skin of the shark, but more
irregular. The sculpturings seem intended evidently
for effect;—to produce harmony of appearance between the scaly coat and the enamelled occipital
plates of bone, the surfaces of the latter are relieved,
where they border on the shoulders, into what seem
scales, just as the dead walls of a building are sometimes, for the sake of uniformity, wrought into blind
windows. The enamelled rays of the fins are finish-

---

deep-sea cockle: the spiny cockle.

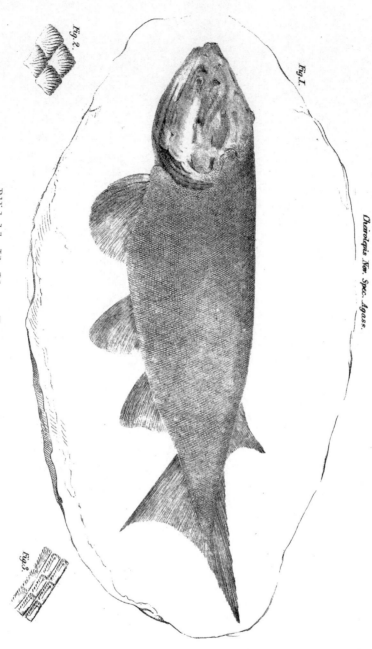

Fig. 2.

Fig. 1.

Fig. 3.

Chæirolepis Nov. Spec. Agass.

PLATE VI

Published by John Johnstone, Hunter Square, Edinburgh.

ed, if I may so speak, after the same style. They lie thick upon one another as the fibres of a quill, and, like these too, they are imbricated on the sides, so that the edge of each seems jagged into a row of prickles. The jaws of the *Cheirolepis* were armed with thickly-set shark-teeth, like those of its cotemporaries the *Osteolepis* and *Diplopterus*.

---

imbricated: this word means overlapping, as with roof tiles, whereas Miller's implied and apparently intended meaning is engaging like, say, the teeth of a modern zip-fastener. However, the **fibres of a quill** (barbs of a feather) do both, thanks to their interlocking barbules, which may be Miller's point.

CHAPTER V.

THERE rests in the neighbourhood of Cromarty, on
the upper stratum of one of the richest ichthyolite
beds I have yet seen, a huge water-rolled boulder of
granitic gneiss, which must have been a traveller in
some of the later periods of geological change, from
a mountain range in the interior highlands of Ross-
shire, more than sixty miles away.   It is an uncouth-
looking mass, several tons in weight, with a flat upper
surface, like that of a table ; and as a table, when engaged in collecting my specimens, I have often found
occasion to employ it.   I have covered it over, times
without number, with fragments of fossil fish,—with
plates, and scales, and jaws, and fins, and, when the
search proved successful, with entire ichthyolites. Why

did I always arrange them, almost without thinking of the matter, into three groupes ?   Why, even when the mind was otherwise employed, did the fragments of the *Coccosteus* and *Ptericthys* come to occupy one corner of the stone, and those of the various fish just described another corner, and the equally well mark- ed remains of a yet different division a third corner ? The process seemed almost mechanical, so little did it employ the attention, and so invariable were the results.   The fossils of the surrounding bed always found their places on the huge stone in three groupes, and at times there was yet a fourth group added,—a group whose organisms belonged not to the animal, but the vegetable kingdom.   What led to the ar- rangement ? or in what did it originate ?   In a prin- ciple inherent in the human mind,—that principle of classification which we find pervading all science,— which gives to each of the many cells of recollec- tion its appropriate facts,—and without which all knowledge would exist as a disorderly and shapeless mass, too huge for the memory to grasp, and too he- terogeneous for the understanding to employ.   I have described but two of the groupes, and must now say a very little about the principle on which, justly or otherwise, I used to separate the third, and on the distinctive differences which rendered the separation so easy.

The recent bony fishes are divided, according to the Cuvierian system of classification, into two great orders, the soft-finned and the thorny-finned order,— the *Malacopterygii* and the *Acanthopterygii*.   In the former the rays of the fins are thin, flexible, articulated, branched ; each ray somewhat resembles a jointed bamboo, with this difference, however, that what seems a single ray at bottom, branches out into three or four rays atop. In the latter (the thorny-finned order),—especially in their anterior dorsal, and perhaps anal fins,—the rays are stiff continuous spikes of bone, and each stands detached as a spear, without joint or branch. The perch may be instanced as a familiar illustration of this order,—the gold-fish of the other. Now, between the fins of two sets—shall I venture to say orders?—of the ichthyolites of the Lower Old Red Sandstone, an equally striking difference obtains. The fin of the *Osteolepis*, with its surface of enamelled and minutely-jointed bones, I have already described as a sort of bird-wing fin. The naked rays, with their flattened surfaces, lay thick together as feathers in the wing of a bird,—so thick as to conceal the connecting membrane ; and fins of similar construction characterized the families of the *Dipterus, Diplopterus, Glyptolepis, Cheiracanthus, Holoptychus*, and, I doubt not, many other families of the same period, which await the researches of future discoverers. But the fins of another set of ichthyolites, their cotemporaries, may be described as batwing fins : they presented to the water a broad expanse of membrane ; and the solitary ray which survives in each was not a jointed, but a continuous spear-like ray. The fins of this set, or order, are thorny fins, like those of the *Acanthopterygii ;* the anterior edge of each, with the exception of perhaps the caudal fin, which differs in construction from the others, is composed of a strong bony spike. Such,

---

Cheiracanthus: should read *Cheirolepis* (author's erratum, emended in 2nd edition).

with some tacit reference, perhaps, to the similar Cuvierian principle of classification, were the distinctive
differences, on the strength of which I used to arrange
two of my groupes of fossils on the granitic boulder;
and the influence of the same principle, almost in-
stinctively exerted,—for in writing the previous pages
I scarce thought of its existence,—has, I find, given to
each group its own chapter.

Of the membranous-finned and thorny-rayed order
of ichthyolites, the *Cheiracanthus*, or thorny-hand, *i. e.*
pectoral, may be regarded as an adequate representative (see Plate VII. fig. 1). The *Cheiracanthus* must
have been an eminently handsome little fish,—slim,
tapering, and described in all its outlines, whether of
the body or the fins, by gracefully-waved lines. It is,
however, a rare matter to find it presenting its original
profile in the stone ;—none of the other ichthyolites
are so frequently distorted as the *Cheiracanthus*. It
seems to have been more a cartilaginous and less an
osseous fish than most of its cotemporaries. However perfect the specimen, no part of the internal
skeleton is ever found, not even when scales as minute as the point of a pin are preserved, and every
spine stands up in its original place. And hence, perhaps, a greater degree of flexibility, and consequent
distortion. The body was covered with small angular scales, brightly enamelled, and delicately fretted
into parallel ridges that run longitudinally along the
upper half of the scale, and leave the posterior portion
of it a smooth glittering surface. They diminish in
size towards the head, which, from the faint stain left
on the stone, seems to have been composed of carti-

---

pectoral: *recte* 'pectoral fin', which Miller had elided, as he did elsewhere and as scien-
tists often do when the context is clear (e.g. 'dorsal' for 'dorsal fin'). In fact, 'pectoral'
means 'pertaining to the shoulder', and 'hand' corresponds to the Greek root *cheir*.

lage exclusively, and either covered with skin or with scales of extreme minuteness. The lower edge of the operculum bears a tagged fringe, like that of a curtain. The tail, a fin of considerable power, had the un-equal-sided character common to the formation; and the slender and numerous rays on both sides are separated by so many articulations as to present the appearance of parallelogrammical scales. The other fins are comparatively of small size. There is a single dorsal placed about two-thirds the entire length of the creature adown the back; and exactly opposite its posterior edge is the anterior edge of the anal fin. The ventral fins are placed high upon the belly, somewhat like those of the perch ; the pectorals only a little higher. But it is rather in the construction of the fins than their position that the peculiarities of the *Cheiracanthus* are most marked. The anterior edge of each, as in the pectorals of the existing genera *Cestracion* and *Chimœra*, is formed of a strong large spine. In the *Chimœra Borealis*, a cartilaginous fish of the Northern Ocean, the spine seems placed in front of the weaker rays, just, if I may be allowed the comparison, as, in a line of mountaineers engaged in crossing a swollen torrent, the strongest man in the party is placed on the upper side of the line, to break off the force of the current from the rest. In the *Cheiracanthus*, however, each fin seems to consist of but a single spine, with an angular membrane fixed to it by one of its sides, and attached to the creature's body on the other. Its fins are masts and sails, the spine representing the mast, and the membrane the sail; and it is a curious characteristic of the order,

---

formation: i.e. this portion of the Old Red Sandstone, and by extension the fishes thereof; **Northern Ocean**: Arctic Ocean, perhaps including the North Atlantic and Barents Sea.

PLATE VII

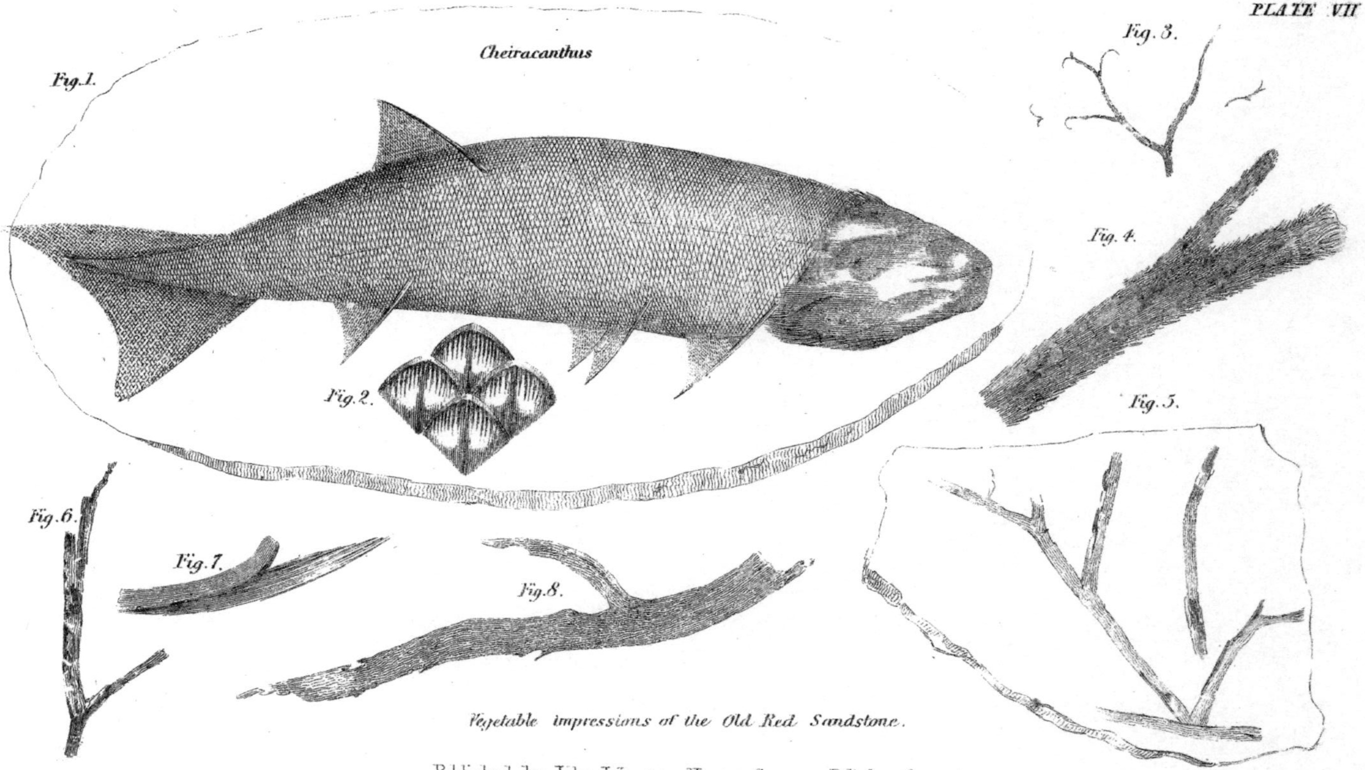

Fig. 3.

Fig. 4.

Fig. 5.

J. Gellatly.

Cheiracanthus

Fig. 1.

Fig. 2.

Fig. 8.

Fig. 7.

Fig. 6.

Vegetable impressions of the Old Red Sandstone.

Published by John Johnstone, Hunter Square, Edinburgh.

that the membrane, like the body, of the ichthyolite, is thickly covered with minute scales. The mouth seems to have opened a very little under the snout, as in the haddock; and there are no indications of its having been furnished with teeth.

An ichthyolite first discovered by the writer about three years ago, and introduced by him to the notice of Agassiz during his recent visit to Edinburgh, but still unfurnished with a name, is a still more striking representative of this order than even the *Cheiracanthus*. It must have been proportionally thick and short, like some of the tropical fishes, though rather handsome than otherwise (see Plate VII. fig. 1). The scales, minute, but considerably larger than those of the *Cheiracanthus*, are of an angular form, and so regularly striated,—the striæ converging to a point at the posterior termination of each scale,—that, when examined with a glass, the body appears as if covered with scallops. It seems a piece of exquisite shell-work, such as we sometimes see on the walls of a grotto. There are two dorsals, the posterior, immediately over the tail, and directly opposite the anal fin; the anterior, somewhat higher up than the ventrals; and all the fins are of great size. The anterior edge of each is formed of a strong spine, round as the handle of a halbert, and diminishing gradually and symmetrically to a sharp point. Though formed externally of solid bone, it seems to have been composed internally of cartilage, like the bones of some of the osseous fishes, those of the halibut for instance; and the place of the cartilage is generally occupied in the stone by carbonate of lime. The

---

recent visit: October 1840; see Appendix 2; **unfurnished with a name**: eventually named *Diplacanthus longispinus* by Agassiz (1844–45); today *Rhadinacanthus* (Appendix 5); **Plate VII**: error for Plate VIII, emended in 2nd edition; **halbert**: halberd, pike (military).

membrane which formed the body of the fin was
covered, like that of the *Cheiracanthus*, with minute
scales, of the same scallop-like pattern with the rest,
but of not more than one-sixth the size of those
which cover the creature's sides and back.   Imagine
two lug-sails stiffly extended between the deck of a
brigantine and her two masts, the latter raking so far
aft as to form an angle of sixty degrees with the horizon, and some idea may be formed of the dorsals of
this singular fish.   They were lug-sails, formed not to
be acted upon by the air, but to act upon the water.
None of my specimens show the head; but judging
from analogies furnished by the other families of the
group, I entertain little doubt that it will be found
to be covered, not by bony plates, but by minute
scales, diminishing, as they approach the snout, into
mere points.   In none of the specimens does any part
of the internal skeleton survive.

My collection contains the remains of yet another
fish of this group, which is still unfurnished with a
name, but which I first discovered about five years
since (see Plate VIII. fig. 2).   It is the smallest ichthyolite of the formation yet known, though not the least
curious.   The length from head to tail in some of
my specimens does not exceed three inches; the largest fall a little short of five.   The scales, which are
of such extreme minuteness that their peculiarities
can be detected by only a powerful glass, resemble
those of the *Cheiracanthus;* but the ridges are more
waved, and seem, instead of running in nearly parallel lines, to converge toward the apex.   There
are two dorsals, the one rising immediately from the

---

lug-sail: an irregularly quadrilateral fore-and-aft sail with its head borne by a yard
loosely fastened to the mast somewhere around the yard's midpoint. It was a common
mainsail rig in small ships and boats. If the mast was **raked** (sloped) back as Miller suggests, the mast and yard would be more or less aligned and seem to be one continuous
spar presenting a triangular (acanthodian) appearance. A brigantine was not usually
rigged in the way described here, and Miller's analogy merges different kinds of rig for
his illustrative purpose; **still unfurnished**: later named *Diplacanthus striatus*, based on
specimens from Miller's collection: Agassiz 1844–45; Andrews 1982a, p. 44.

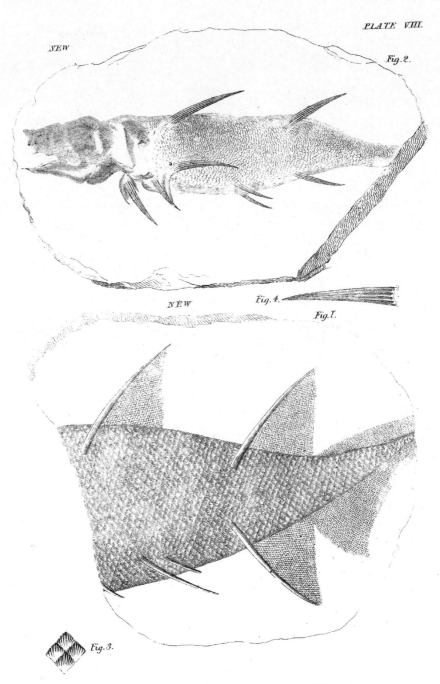

PLATE VIII.

NEW

Fig. 2.

NEW

Fig. 4.

Fig. 1.

Fig. 3.

Published by John Johnstone, Hunter Square, Edinburgh.

shoulder, a little below the nape; the other directly opposite the anal fin. The ventrals are placed near the middle of the belly. There is a curious mechanism of shoulder-bone involved with a lateral spine and with the pectorals. The creature, unlike the *Cheiracanthus*, seems to have been furnished with jaws of bone: there are fragments of bone upon the head, tuberculated apparently on the outer surface; and minute cylinders of carbonate of lime running along all the larger bones, where we find them accidentally laid open, show that they were formed on internal bases of cartilage. But the best-marked characteristic of the creature is furnished by the spines of its fins, which are of singular beauty. Each spine resembles a bundle of rods, or rather, —like a Gothic column,—the sculptured semblance of a bundle of rods, which finely diminish towards a point, sharp and tapering as that of a rush. The rest of the fin presents the appearance of a mere scaly membrane, and no part of the internal skeleton appears. Perhaps this last circumstance, common to all the ichthyolites of the formation, if we except the families of the *Coccosteus* and *Pterichthys*, may throw some light on the apparently membranous condition of fin peculiar to the families of this order. What appears in the fossil a mere scaly membrane attached to a single spine of bone, may have had in the living animal a cartilaginous frame-work, like the fins of the dog-fish and thorn-back, that are amply furnished with rays of cartilage,—though, of course, all such rays must have disappeared in the stone, like the rest of the internal skeleton. Unquestionably the

---

Gothic column: clustered columns or cluster-columns are found in many medieval cathedrals, sometimes resembling a **bundle of rods** as in the transept crossing at Salisbury.

caudal fin of the two unnamed fossils described must have been strengthened by some such internal frame-work; for, as they differ from the other fins in being unprovided with osseous spines, they would have formed, without an internal skeleton, mere pendu-lous attachments, altogether unfitted to serve the purpose of instruments of motion. There may be found in the bony spines of all this order direct proof that had there been an internal skeleton of bone, it would have survived. The spines run deep into the body, as a ship's masts run deep into her hulk; and we can see them standing up among the scales to their termination, in such bold relief, that, from a sort of pictorial illusion, they seem as if fixed to the creature's sides, and foreshortened, instead of rising in profile from its back or belly (see Plate VIII. fig 1). The observer will of course remember, that in the living animal the view of the spine must have termi-nated with the line of the profile, just as the view of a vessel's mast terminates with the deck, though the mast itself penetrates to the interior keel. Now it must be deemed equally obvious, that had the ver-tebral column been of bone, not of cartilage, instead of exhibiting no trace, even the faintest, of having ever existed, it would have stood out in as high re-lief as the internal butts or stocks of the spines. And such are the general characteristics of a few of the ichthyolites of this lower formation of the Old Red Sandstone,—a few of the more striking forms, sculp-tured, if I may so speak, on the middle compartment of the Caithness pyramid. It would be easy render-ing the list more complete at even the present stage,

---

hulk: hull; lower formation ... middle compartment: in the Caithness pyramid on pp. 30–33.

when the field is still so new that almost every labourer in it can exhibit genera and species unknown
to his brother-labourers. The remains of a species of
Holoptychius have been discovered low in the formation, at Orkney, by Dr Traill; similar remains have
been found in it at Gamrie. In its upper beds the
specimens seem so different from those in the lower,
that in extensive collections made from the inferior
strata of one locality, Agassiz has been unable to identify a single specimen with the specimens of collections made from the superior strata of another,
though the genera are the same. Meanwhile there
are heads and hands at work on the subject; Geology has become a Briareus; and I have little doubt
that in five years hence this third portion of the Old
Red Sandstone will be found to contain as many distinct varieties of fossil fish as the whole geological
scale was known to contain fifteen years ago.

There is something very admirable in the consistency of style which obtains among the ichthyolites
of this formation. In no single fish of either group
do we find two styles of ornament,—in scarce any two
fishes do we find exactly the same style. I pass fine
buildings almost every day. In some there is a dis-
cordant jumbling,—an Egyptian Sphynx, for instance,
placed over a Doric portico; in all there prevails a
vast amount of timid imitation. The one repeats
the other, either in general outline or in the subordinate parts. But the case is otherwise among the
ichthyolites of the Old Red Sandstone; nor does it
lessen the wonder, that their nicer ornaments should
yield their beauty only to the microscope. There is

Agassiz ... same: Miller is referring to the 'formation' in the sense of the lower part of
the Old Red Sandstone system, as used in the previous page, and noting that the fishes
of some localities ascribed to the lower part of the formation are different from those
higher up in this same formation. The reference is probably to Agassiz's spoken comments at Altyre and Edinburgh (pp. 134–35); **Briareus**: one of the hundred-handed
giants of Greek mythology, said to have fifty heads in some Classical sources (e.g. the
Library of Pseudo-Apollodorus); **nicer**: more delicate.

unity of character in every scale, plate, and fin,—
unity such as all men of taste have learned to ad-
mire in those three Grecian orders from which the ingenuity of Rome was content to borrow, when it pro-
fessed to invent,—in the masculine Doric,—the chaste
and graceful Ionic,—the exquisitely elegant Corin-
thian; and yet the unassisted eye fails to discover
the finer evidences of this unity: it would seem as
if the adorable Architect had wrought it out in secret
with reference to the Divine idea alone. The artist
who sculptured the cherry-stone consigned it to a ca-
binet, and placed a microscope beside it;—the microscopic beauty of these ancient fish was consigned
to the twilight depths of a primeval ocean. There is
a feeling which at times grows upon the painter and
the statuary, as if the love and the perception of the
beautiful had been sublimed into a kind of moral
sense. Art comes to be pursued for its own sake; the
exquisite conception in the mind, or the elegant and
elaborate model, becomes all in all to the worker, and
the dread of criticism or the appetite of praise almost nothing. And thus, through the influence of
a power somewhat akin to conscience, but whose province is not the just and the good, but the fair, the
refined, the exquisite, have works prosecuted in solitude, and never intended for the world, been found
fraught with loveliness. Sir Thomas Lawrence, when
finishing, with the most consummate care, a picture
intended for a semi-barbarous foreign court, was asked why he took so much pains with a piece destined, perhaps, never to come under the eye of a
connoisseur. " I cannot help it," he replied; " I do

---

orders: of Classical architecture; adorable: worthy of worship; cherry-stone: Miller
may have been thinking of the celebrated 'Cherry Stone with 185 Carved Faces' sculpted
in the sixteenth century and now housed in the Grünes Gewölbe, Dresden; microscope: probably in the sense of a magnifying glass (see note for p. 55); Lawrence: the
celebrated English portrait painter (1769–1830) expressed similar views of his portrait of
the Persian (i.e. Iranian) ambassador to the UK, Mirza Abolhassan (1810): Williams
1831, vol. II, pp. 52–53. Miller's version has a very different tone and no verbal echoes,
and may be an unconscious reworking or derived from an untraced printed source.

the best I can, unable, through a tyrant feeling, that will not brook offence, to do any thing less." It would be perhaps over-bold to attribute any such o'ermastering feeling to the Creator; and yet certain it is, that among his creatures well nigh all approximations towards perfection, in the province in which it expatiates, owe their origin to it, and that Deity in all his works is his own rule.

The *Osteolepis* was cased, I have said, from head to tail in complete armour. The head had its plaited mail, the body its scaly mail, the fins their mail of parallel and jointed bars; the entire suit glittered with enamel; and every plate, bar, and scale, was dotted with microscopic points. Every ray had its double or treble punctulated row, every scale or plate its punctulated group; the markings lie as thickly in proportion to the fields they cover as the circular perforations in a lace veil; and the effect viewed through the glass is one of lightness and beauty. In the *Cheirolepis* an entirely different style obtains. The enamelled scales and plates glitter with minute ridges, that show like thorns in a December morning varnished with ice. Every ray of the fins presents its serrated edge, every occipital plate and bone its sculptured prominences, every scale its bunch of prickle-like ridges. A more rustic style characterized the *Glyptolepis*. The enamel of the scales and plates is less bright; the sculpturings are executed on a larger scale, and more rudely finished. The relieved ridges, waved enough to give them a pendulous appearance, drop adown the head and body. The rays of the fins, of great length, present also a pendulous

G

---

expatiates: in the old sense of 'ranges, moves freely' ('it', five words later, refers to the o'ermastering feeling); rustic: crude; relieved: raised, brought into relief.

appearance. The bones and scales seem disproportionately large. There is a general rudeness in the finish of the creature, if I may so speak, that reminds one of the tatooings of a savage, or the corresponding style of art in which he ornaments the handle of his stone-hatchet or his war-club. In the *Cheiracanthus*, on the contrary, there is much of a minute and cabinet-like elegance. The silvery smoothness of the fins, dotted with scarcely visible scales, harmonized with a similar appearance of head; a style of sculpture resembling the parallel etchings of the line engraver fretted the scales; the fins were small, and the contour elegant. I have already described the appearance of the unnamed fossils,—the seeming shellwork that covered the sides of the one,—its mast-like spines and sail-like fins; and the Gothic-like peculiarities that characterized the other,—its rodded, obelisk-like spires, and the external framework of bone that stretched along its pectorals.

Till very lately it was held that the Old Red Sandstone of Scotland contained no mollusca. It seemed difficult, however, to imagine a sea abounding in fish, and yet devoid of shells. In all my explorations, therefore, I had an eye to the discovery of the latter, and on two several occasions I disinterred what I supposed might have formed portions of a cardium or terebratula. On applying the glass, however, the punctulated character of the surface showed that the supposed shells were but parts of the concave helmet-like plate that covered the snout of the *Osteolepis*. In the ichthyolite beds of Cromarty and Ross, of Moray, Banff, Perth, Forfar, Fife, and Berwickshire, not a single shell has yet been found; but there have been discovered of late, in the upper beds of the Lower Old Red Sandstone in Caithness and Orkney, the remains of a small delicate bivalve, not yet described or figured, but which very much resembles a *Venus*. In the Tilestones of England, so carefully described by Mr Murchison in his *Silurian System*, shells are very abundant; and the fact may now be regarded as established, that the Tilestones of England belong to a deposit contemporaneous with the ichthyolite beds of Caithness and Cromarty. They occupy the same place low in the base of the Old Red; and there is at least one ichthyolite common to both,* and which does not occur in the superior strata of the system in either country, — the *Dipterus Macrolepidotus*. The evidence that the fish and shells lived in the same period, and represent therefore the same formation, may be summed up in a single sentence. We learn from the geology of Caithness that this species of *Dipterus* was unquestionably cotemporary with all the other ichthyolites described;—we learn from the geology of Herefordshire that the shells were as unquestionably cotemporary with it. These—the shells —are of a singularly mixed character, regarded as a group, uniting, says Mr Murchison, forms at one time deemed characteristic of the more modern formations, —of the later secondary, and even tertiary periods,— with forms the most ancient, and which characterize the molluscous remains of the transition rocks. Turbinated shells and bivalves of well nigh the recent type

* *Silurian System*, part ii. page 599.

bivalve: see also p.ix.

may be found lying side by side with chambered Orthoceratites and Terebratula.*

The vegetable remains of the formation are numerous but obscure, consisting mostly of carbonaceous markings, such as might be formed by comminuted sea-weed. (See Plate VII.) Some of the impressions fork into branches at acute angles; some affect a waved outline ; most of them, however, are straight and undivided. They lie in some places so thickly in layers as to give the stone in which they occur a slaty character. One of my specimens shows minute markings, somewhat resembling the bird-like eyes of the Stigmaria Ficoides of the Coal Measures ;—the branches of another terminate in minute hooks, that remind one of the hooks of the young tendrils of the pea when they first begin to turn. In yet another there are marks of the ligneous fibre : when examined by the glass it resembles a bundle of horse-hairs lying stretched in parallel lines ; and in this specimen alone have I found aught approaching to proof of a terrestrial origin. The deposition seems to have taken place far from land ; and this lignite, if in reality such, had probably drifted far ere it at length became weightier than the supporting fluid, and sank. It is by no means rare to find fragments of wood that have been borne out to sea by the gulf-stream from the shores of Mexico or the West Indian Islands, stranded on the rocky coasts of Orkney and Shetland.

The dissimilarity which obtains between the fossils of the cotemporary formations of this system in Eng-

---

* *Silurian System,* part i. p. 183.

Stigmaria Ficoides: a large fossil plant, whose surface is dotted with **bird-like eyes**; see also p. 271.

land and Scotland is instructive. The group in the one consists mainly of molluscous animals,—in the other almost entirely of ichthyolites, and what seems to have been algæ. Other localities may present us with yet different groupes of the same period,—with the productions of its coasts, its lakes, and its rivers. At present we are but beginning to know just a little of its littoral shells, and of the fish of its profounder depths. These last are surely curious subjects of inquiry. We cannot catechize our stony ichthyolites, as the necromantic lady of the *Arabian Nights* did the coloured fish of the lake, which had once been a city, when she touched their dead bodies with her wand, and they straightway raised their heads and replied to her queries. We would have many a question to ask them if we could,—questions never to be solved. But even the contemplation of their remains is a powerful stimulant to thought. The wonders of Geology exercise every faculty of the mind,—reason, memory, imagination; and though we cannot put our fossils to the question, it is something to be so aroused as to be made to put questions to one's self. I have referred to the consistency of style which obtained among these ancient fishes,—the unity of character which marked every scale, plate, and fin of every various family, and which distinguished it from the rest. And who can doubt that the same shades of variety existed in their habits and their instincts? We speak of the infinity of Deity,—of His inexhaustible variety of mind; but we speak of it until the idea becomes a piece of mere common-place in our mouths. It is well to be brought to feel, if

---

catechize: interrogate. Especially in Scottish Presbyterianism, a catechist is a person, usually an elder, who instructs and tests young people in Church doctrine, often by asking questions; *Arabian Nights*: Miller here (and perhaps also on p. 60) alludes to the final third of the story of 'The Fisherman and the *'Ifrit*', in which a city and its inhabitants are magically turned into a lake of coloured fish. When caught and fried, the fishes are still capable of chanting to the sorceress who addresses them: Nights 6–9 in Lyons 2008, vol. I, pp. 34–50.

not to conceive of it,—to be made to know that we ourselves are barren-minded, and that in Him " all fulness dwelleth." Succeeding creations, each with its myriads of existences, do not exhaust Him. He never repeats himself. The curtain drops at his command over one scene of existence full of wisdom and beauty, —it rises again, and all is glorious, wise, and beautiful as before, and all is new. Who can sum up the amount of wisdom whose record He has written in the rocks,—wisdom exhibited in the succeeding creations of earth, ere man was, but which was exhibited surely not in vain. May we not say with Milton,—

> Think not though men were none,
> That heaven could want spectators, God want praise;
> Millions of spiritual creatures walked the earth,
> And these with ceaseless praise his works beheld.

It is well to return on the record, and to read in its unequivocal characters the lessons which it was intended to teach. Infidelity has often misinterpreted its meaning, but not the less on that account has it been inscribed for purposes alike wise and benevolent. Is it nothing to be taught with a demonstrative evidence which the metaphysician cannot supply, that races are not eternal,—that every family had its beginning, and that whole creations have come to an end ?

---

all fulness: Bible, Colossians 2:9, 'For in Him [Christ] dwelleth all the fulness of the Godhead bodily'; Think not ... : lines slightly condensed from Adam's observations on his angelic companions in Milton, *Paradise Lost*, Book IV, lines 675–79, adapted to fit both the past-tense context of Miller's suggestion and the syntax which frames the quotation; races: here meaning species, families and orders, not varieties within a species (compare pp. 240, 275).

CHAPTER VI.

The Lines of the Geographer rarely right Lines.—These last, however, always worth looking at when they occur.—Striking instance in the Line of the Great Caledonian Valley.—Indicative of the Direction in which the Volcanic Agencies have operated.—Sections of the Old Red Sandstone furnished by the Granitic Eminences of the Line.—Illustration.—Lias of the Moray Frith.—Surmisings regarding its original Extent.—These lead to an Exploratory Ramble.—Narrative.—Phenomena exhibited in the course of half an hour's Walk.—The little Bay.—Its Strata and their Organisms.

THE natural boundaries of the geographer are rarely described by right lines. Wherever these occur, however, the geologist may look for something remarkable. There is one very striking example furnished by the north of Scotland. The reader, in consulting a map of the kingdom, will find that the edge of a ruler laid athwart the country in a direction from south-west to north-east, touches the whole northern side of the great Caledonian Valley, with its long straight line of lakes,—and onwards, beyond the valley's termination at both ends, the whole northern side of Loch Eil and Loch Linnhe, and the whole of the abrupt and precipitous northern shores of the Moray Frith, to the extreme point of Tarbat Ness,—a right line of

---

right lines: straight lines; **Caledonian Valley**: the Great Glen, with Lochs Lochy, Oich, and Ness; see Figure 13.

considerably more than a hundred miles. Nor does
the geography of the globe furnish a line better defined by natural marks. There is both rampart and
fosse. On the one hand we have the rectilinear lochs
and lakes, with an average profundity of depth more
than equal to that of the German Ocean, and, added
to these, the rectilinear lines of frith ; on the other
hand, with but few interruptions, there is an inclined
wall of rock, which rises at a steep angle in the interior to nearly two thousand feet over the level of the
Great Canal, and overhangs the sea towards its northern termination, in precipices of more than a hundred
yards.

The direction of this rampart and fosse—this Roman wall of Scottish geological history—seems to
have been that in which the volcanic agencies chiefly
operated in upheaving the entire island from the
abyss. The line survives as a sort of foot-track, hollowed by the frequent tread of earthquakes, to mark
the course in which they journeyed. Like one of the
great lines in a trigonometrical survey, it enables us,
too, to describe the lesser lines, and to determine their
average bearing. *The volcanic agencies must have
extended athwart the country from south-west to north-
east.* Mark in a map of the island—all the better if
it be a geological one—the line in which most of our
mountain ranges stretch across from the German
Ocean to the Atlantic,—the line, too, in which our
friths, lochs, and bays, on both the eastern and western coasts, and especially those of the latter, run into
the interior Mark, also, the line of the geological
formations, where least broken by insulated groupes

---

fosse: defensive ditch; **Great Canal**: Telford's Caledonian Canal along the Great Glen
from Inverness to near Fort William; **Roman wall**: Hadrian's Wall or the Antonine
Wall, different northern boundaries of the Roman defensive zone in northern Britain;
great lines in a trigonometrical survey: one of the primary transects across country
from which secondary triangulation was done to fill in the map.

of hills,—the line, for instance, of the Old Red Sand-stone belt, which flanks the southern base of the Grampians,—the nearly parallel line of our Scottish Coal-field, in its course from sea to sea,—the line of the Grauwacke, which forms so large a portion of the south of Scotland,—the line of the English Coal-field, of the Lias, of the Oolite, of the Chalk,—and how in this process of diagonal lining, if I may so speak, the south-eastern portion of England comes to be cut off from the secondary formations altogether, and, but for the denudation of the valley of the Weald, would have exhibited only tertiary depositions. In all these lines, whether of mountains, lakes, friths, or formations, there is an approximation to parallelism with the line of the great Caledonian Valley,—proofs that the upheaving agency from beneath must have acted in this direction from some unknown cause, dur-ing all the immensely-extended term of its operations, and along the entire length of the island. It is a fact not unworthy of remark, that the profound depths of Loch Ness undulated in strange sympathy with the reeling towers and crashing walls of Lisbon, dur-ing the great earthquake of 1755; and that the im-pulse, true to its ancient direction, sent the waves in huge furrows to the north-east and the south-west.

The north-eastern portion of this rectilinear wall or chain runs, for about thirty miles, through an Old Red Sandstone district. The materials which com-pose it are as unlike those of the plain out of which it arises, as the materials of a stone-dike running half-way into a field are unlike the vegetable mould which forms the field's surface. The ridge itself is

---

exhibited only tertiary depositions: would only show strata of Tertiary age; in fact erosion cut down through them into the Chalk and Greensand; wall: the 'Roman wall' of the Great Glen and its extensions, p. 104; Old Red Sandstone district: including the Black Isle and the Sutors of Cromarty; dike: in the sense of a boundary wall (as in Offa's Dyke, Scots' Dike), not a ditch.

of a granitic texture,—a true gneiss. At its base we find only conglomerates, sandstones, shales, and stratified clays, and these lying against it in very high angles. Hence the geological interest of this lower portion of the wall. As has been shrewdly remarked by Mr Murchison,* in one of his earlier papers, the gneiss seems to have been forced through the sandstone from beneath, in a solid, not a fluid form; and as the ridge atop is a narrow one, and the sides remarkably abrupt,—an excellent wedge both in consistency and form,—instead of having acted on the surrounding depositions, as most of the south-country traps have done that have merely issued from a vent, and overlaid the upper strata, it has torn up the entire formation from the very bottom. Imagine a large wedge forced from below through a sheet of thick ice on a river or pond. First the ice rises in an angle, that becomes sharper and higher as the wedge rises; then it cracks and opens, presenting its upturned edges on both sides, and through comes the wedge. And this is a very different process, be it observed, from what takes place when the ice merely cracks, and the water issues through the crack. In the one case there is a rent, and water diffused over the surface;—in the other there is the projecting wedge, flanked by the upturned edges of the ice; and these edges of course serve as indices to decide regarding the ice's thickness, and the various layers of which it is composed. Now, such are the phenomena exhibited by the wedge-like granitic ridge. The Lower

" See Transactions of the London Geological Society for 1828, p. 354.

Murchison: the paper cited, in fact, leaves open the question of whether the Sutors gneiss was solid or liquid when it was uplifted (Murchison 1828, pp. 353–56).

Old Red Sandstone, tilted up against it on both sides at an angle of about eighty, exhibits in some parts a section of well nigh two thousand feet, stretching from the lower conglomerate to the soft unfossiliferous sandstone, which forms in Ross and Cromarty the upper beds of the formation. There is a mighty advantage to the geologist in this arrangement. When books are packed up in a deep box or chest, we have to raise the upper tier ere we can see the tier below, and this second tier ere we can arrive at a third, and so on to the bottom. But when well arranged on the shelves of a library, we have merely to run the eye along their lettered backs, and we can thus form an acquaintance with them at a glance, which in the other case would have cost us a good deal of trouble. Now, in the neighbourhood of this granitic wedge, or wall, the strata are arranged, not like books in a box,—such was their original position,—but like books on the shelves of a library. They have been unpacked and arranged by the uptilting agent; and the knowledge of them, which could only have been attained in their first circumstances by perforating them with a shaft of immense depth, may now be acquired simply by passing over their edges. A morning's saunter gives us what would have cost, but for the upheaving granite, the labour of a hundred miners for five years.

By far the greater portion of the life of the writer was spent within less than half an hour's walk of one of these upturned edges. I have described the granitic rock, with reference to the disturbance it has occasioned, as a wedge forced from below, and with reference to

---

eighty: degrees to the horizontal (compare p. 115); **spent ... walk**: at Cromarty to

its rectilinear position in the sandstone district which
it traverses, as a stone-wall running half-way into a
field. It may communicate a still correcter and
livelier idea to think of it as a row of wedges, such as
one sometimes sees in a quarry when the workmen
are engaged in cutting out from the mass some immense block, intended to form a stately column or
huge architrave. The eminences, like the wedges,
are separated—in some places the sandstone lies between,—in others there occur huge chasms filled by
the sea. The Friths of Cromarty and Beauly, for instance, and the Bay of Munlochy, open into the interior between these wedge-like eminences ;—the
well known Sutors of Cromarty represent two of the
wedges, and it was the section furnished by the Southern Sutor that lay so immediately in the writer's
neighbourhood. The line of the Cromarty Frith
forms an angle of about thirty-five degrees with that
of the granitic line of wedge-like hills which it bisects ; and hence the peculiar shape of that tongue of
land which forms the lower portion of the Black Isle,
and which, washed by the Moray Frith on the one
side, and by the Frith of Cromarty on the other, has
its apex occupied by the Southern Sutor. Imagine a
lofty promontory somewhat resembling a huge spear
thrust horizontally into the sea,—a ponderous mass
of granitic gneiss, of about a mile in length, forming
the head, and a rectilinear line of the Old Red Sandstone, more than ten miles in length, forming the
shaft ; and such is the appearance which this tongue
of land presents when viewed from its north-western
boundary, the Cromarty Frith. When viewed from

the Moray Frith,—its south-western boundary,—we see the same granitic spear-head, but find the line of the shaft knobbed by the other granitic eminences of the chain.

Now on this tongue of land I first broke ground as a geologist. The quarry described in my introductory chapter, as that in which my notice was first attracted by the ripple-markings, opens on the Cromarty-Frith side of this huge spear-shaft; the quarry to which I removed immediately after, and beside which I found the fossils of the Lias, opens on its Moray-Frith side. The uptilted section of sandstone occurs on both sides, where the shaft joins to the granitic spear-head, but the Lias I found on the Moray-Frith side alone. It studs the coast in detached patches, sorely worn by the incessant lashings of the Frith, and each patch bears an evident relation, in the place it occupies, to a corresponding knob or wedge in the granitic line. The Northern Sutor, as has been just said, is one of these knobs or wedges. It has its accompanying patch of Lias upheaved at its base, and lying unconformably, not only to its granitic strata, but also to its subordinate sandstones. The Southern Sutor, another of these knobs, has also its accompanying patch of Lias, which, though lying beyond the fall of the tide, strews the beach, after every storm from the east, with its shales and its fossils. The hill of Eathie is yet another knob of the series, and it, too, has its Lias patch. The granitic wedges have not only uptilted the sandstone, but they have also upheaved the superincumbent Lias, which, but for their agency, would have remained buried under the waters of the Frith,

---

tongue of land: the north end of the Black Isle, including Cromarty burgh.

and its ever-accumulating banks of sand and gravel.
I had remarked at an early period the correspondence
of the granitic knobs with the Lias patches, and
striven to realise the original place and position of the
latter ere the disturbing agent had upcast them to the
light. What, I have asked, was the extent of this
comparatively modern formation in this part of the
world, ere the line of wedges were forced through
from below? A wedge struck through the ice of a
pond towards the centre breaks its continuity, and we
find the ice on both sides the wedge ; whereas, when
struck through at the pond edge, it merely raises the
ice from the bank, and we find it, in consequence,
on but one side the wedge. Whether, have I often
inquired, were the granitic wedges of this line forced
through the Lias at one of its edges, or at a comparatively central point ? and about ten years ago I set
myself to ascertain whether I could not solve the
question. The Southern Sutor is a wedge open to
examination on both its sides ;—the Moray Frith
washes it upon the one side, the Cromarty Frith on
the other. Was the Lias to be found on both its
sides ? If so, the wedge must have been forced *through*
the formation, not merely *beside* it. It occurs, as I
have said, on the Moray-Frith side the wedge; and
I resolved on carefully exploring the Frith of Cromarty, to try whether it did not occur on that side
too.

With this object I set out on an exploratory excursion, on a delightful morning of August 1830.
The tide was falling ; it had already reached the
line of half-ebb; and from the Southern Sutor to

the low, long promontory on which the town of Cromarty is built, there extended a broad belt of mingled sand-banks and pools, accumulations of boulders and shingle, and large tracks darkened with algæ. I passed direct by a grassy pathway to the Sutor,— the granitic spear-head of a late illustration,—and turned, when I reached the curved and contorted gneiss, to trace through the broad belt left by the retiring waters, and in a line parallel to what I have described as the shaft of the huge spear, the beds and strata of the Old Red Sandstone in their ascending succession. I first crossed the conglomerate base of the system, here little more than a hundred feet in thickness. The ceaseless dash of the waves, which smooth most other rocks, has a contrary effect on this bed, except in a few localities, where its arenaceous cement or base is much indurated. Under both the Northern and Southern Sutors the softer cement yields to the incessant action, while the harder pebbles stand out in bold relief; so that wherever it presents a mural front to the breakers, we are reminded, by its appearance, of the artificial rock-work of the architect. It roughens as the rocks around it polish. Quitting the conglomerate, I next passed over a thick bed of coarse red and yellowish sandstone, with here and there a few pebbles sticking from its surface, and here and there a stratum of finer-grained fissile sandstone inserted between the rougher strata; I then crossed over strata of an impure grayish limestone and a slaty clay, abounding, as I long afterwards ascertained, in ichthyolites and vegetable remains. There are minute veins in the limestone (apparently cracks filled

----

tracks: tracts; **late illustration**: p. 109.

up), of a jet black bituminous substance resembling anthracite; the stratified clay is mottled by layers of semi-
aluminous semi-calcareous nodules, arranged like layers of flint in the upper Chalk.   These nodules, when
cut up and polished, present very agreeable combinations of colour; there is generally an outer ring of reddish brown, an inner ring of pale yellow, and a central
patch of red, and the whole is prettily veined with
dark-coloured carbonate of lime.*   Passing onwards
and upwards in the line of the strata, I next crossed
over a series of alternate beds of coarse sandstone and
stratified clay, and then lost sight of the rock altogether, in a wide waste of shingle and boulder-stones,
resting on a dark blue argillaceous diluvium, sometimes employed in that part of the country, from its
tenacious and impermable character, for lining ponds
and dams, and as mortar for the foundations of
low-lying houses exposed in wet weather to the sudden rise of water.   The numerous boulders of this tract
have their story to tell, and it is a curious one.   The
Southern Sutor, with its multitudinous fragments of
gneiss, torn from its sides by the sea, or loosened by
the action of frosts and storms, and rolled down its
precipices, is only a few hundred yards away;—its
base, where these lie thickest, has been swept by tempests, chiefly from the east, for thousands and thousands of years; and the direct effect of these tempests,
regarded as transporting agents, would have been to
strew this stony tract with those detached fragments.

* A concretionary limestone of the Old Red system in England, variegated with purple and green, was at one time
wrought as a marble. (*Silurian System*, part i. p. 176.)

like layers of flint: occurring at certain specific levels or bands within the matrix of
stratified clay; concretionary limestone: in south Herefordshire (Murchison 1839, part I,
p. 177).

The same billow that sends its long roll from the German Ocean to sweep the base of the Sutor, and to leap up against its precipices to the height of eighty and a hundred feet, breaks in foam, only a minute after, over this stony tract; which has, in consequence, its sprinkling of fragments of gneiss transported by an agency so obvious. But for every one such fragment which it bears, we find at least ten boulders that have been borne for forty and fifty miles in the opposite direction from the interior of the country,—a direction in which no transporting agency now exists. The tempests of thousands of years have conveyed for but a few hundred yards not more than a tithe of the materials of this tract; nine-tenths of the whole have been conveyed by an older agency over spaces of forty and fifty miles. How immensely powerful, then, or how immensely protracted in its operation, must that older agency have been !

I passed onwards, and reached a little bay, or rather angular indentation of the coast, in the neighbourhood of the town. It was laid bare by the tide this morning far beyond its outer opening; and the huge table-like boulder, which occupies nearly its centre, and to which, in a former chapter, I have had occasion to refer, held but a middle place between the still darkened flood-line that ran high along the beach, and the brown line of ebb that bristled far below with forests of the rough-stemmed tangle. This little bay or inflection of the coast serves as a sort of natural wear in detaining floating driftweed, and is often found piled, after violent storms from

H

---

wear: weir.

the east, with accumulations, many yards in extent
and several feet in depth, of kelp and tangle, mixed
with zoophites and mollusca, and the remains of fish
killed among the shallows by the tempest. Early
in the last century, a large body of herrings, pursued
by whales and porpoises, were stranded in it, to the
amount of several hundred barrels; and it is said that
salt and cask failed the packers when but comparatively a small portion of the shoal were cured, and
that by much the greater part of them were carried
away by the neighbouring farmers for manure.   Ever
since the formation of the present coast-line, this natural wear has been arresting, tide after tide, its heaps
of organic matter, but the circumstances favourable
to their preservation have been wanting: they ferment and decay when driven high on the beach; and
the next spring-tide, accompanied by a gale from the
west, sweeps every vestige of them away; and so,
after the lapse of many centuries, we find no other
organisms among the rounded pebbles that form the
beaches of this little bay, than merely a few broken
shells, and occasionally a mouldering fish-bone. Thus
very barren formations may belong to periods singularly rich in organic existences. When what is now
the little bay was the bottom of a profound ocean,
and far from any shore, the circumstances for the
preservation of its organisms must have been much
more favourable.   In no locality in the Old Red
Sandstone with which I am acquainted, have such
beautifully-preserved fossils been found. But I anticipate.

In the middle of the little bay, and throughout the

---

formation of the present coast-line: a geologically recent event. To modern geologists,
as to Miller, the flat ground behind the beach, now called Reeds Park, is an excellent
example of a raised beach with a well-marked if now somewhat degraded fossil sea-cliff
behind.

greater part of its area, I found the rock exposed,—a circumstance which I had marked many years before when a mere boy, without afterwards recurring to it as one of interest. But I had now learned to look at rocks with another eye ; and the thought which first suggested itself to me regarding the rock of the little bay was, that I had found the especial object of my search—the Lias. The appearances are in some respects not dissimilar. The Lias of the north of Scotland is represented in some localities by dark-coloured unctuous clays, in others by grayish-black sandstones that look like indurated mud, and in others by beds of black fissile shale, alternating with bands of coarse impure limestone, and studded between the bands with limestone nodules of richer quality and finer grain. The rock laid bare in the little bay is a stratified clay, of a gray colour tinged with olive, and occurring in beds separated by indurated bands of gray micaceous sandstone. They also abound in calcareous nodules. The dip of the strata, too, is very different from that of the beds which lean against the gneiss of the Sutor. Instead of an angle of eighty, it presents an angle of less than eight. The rocks of the little bay must have lain beyond the disturbing uptilting influence of the granitic wedge. So thickly are the nodules spread over the surface of some of the beds, that they reminded me of floats of broken ice on the windward side of a lake after a few days' thaw, when the edges of the fragments are smoothed and rounded, and they press upon one another, so as to cover, except in the angular interstices, the entire surface.

I set myself carefully to examine. The first nodule I laid open contained a bituminous-looking mass, in which I could trace a few pointed bones and a few minute scales. The next abounded in rhomboidal and finely-enamelled scales, of a much larger size and more distinct character. I wrought on with the eagerness of a discoverer entering for the first time in a *terra incognita* of wonders. Almost every fragment of clay, every splinter of sandstone, every limestone nodule, contained its organism. Scales, spines, plates, bones, entire fish ; but not one organism of the Lias could I find,—no ammonites, no belemnites, no gryphites, no shells of any kind; the vegetable impressions were entirely different ; and not a single scale, plate, or ichthyodorulite could I identify with those of the newer formation. I had got into a different world, and among the remains of a different creation; but where was its proper place in the scale? The beds of the little bay are encircled by thick accumulations of diluvium and debris, nor could I trace their relation to a single known rock. I was struck, as I well might, by the utter strangeness of the forms,—the oar-like arms of the *Pterichthys* and its tortoise-like plates,—the strange buckler-looking head of the *Coccosteus*, which, I supposed, might possibly be the back of a small tortoise, though the tubercles reminded me rather of the skin of the shark,—the polished scales and plates of the *Osteolepis*,—the spined and scaled fins of the *Cheiracanthus*,—above all, the one-sided tail of at least eight out of the ten or twelve varieties of fossil which the deposit contained. All together excited and astonished me. But some time elapsed ere I learned to distinguish the nicer generic differences of the various organisms of the formation. I found fragments of the *Ptericthys* on this first morning; but I date its discovery in relation to the mind of the discoverer, more than a twelvemonth later.* I confounded the *Cheiracanthus*, too, with its single-spined and membranous dorsal, with one of the still unnamed fossils, furnished with two such dorsals; and the *Diplopterus* with the *Osteolepis.* Still, however, I saw enough to exhilarate and interest; I wrought on till the advancing tide came splashing over the nodules, and a powerful August sun had risen towards the middle sky; and were I to sum up all my happier hours, the hour would not be forgotten in which I sat down on a rounded boulder of granite, by the edge of the sea, when the last bed was covered, and spread out on the beach before me the spoils of the morning.

* I find by some notes which had escaped my notice when drawing up for the *Witness* newspaper the sketches now expanding into a volume, that in the year 1834 I furnished the collection of a geological friend, the Rev. John Swanson, minister of the parish of Small-Isles in the Outer Hebrides, with a well-marked specimen of the *Ptericthys Milleri.* The circumstance pleasingly reminds me of the first of all my early acquaintance, who learned to deem the time not idly squandered that was spent in exploring the wonders of by-gone creations. Does the minister of Small-Isles still remember the boy who led him in quest of petrifactions,—himself a little boy at the time,—to a deep solitary cave on the Moray Frith, where they lingered amid stalactites and mosses, till the wild sea had surrounded them unmarked, barring all chance of retreat, and the dark night came on?

Swanson: Rev. John Swanson (1804–74), Miller's boyhood friend, and (in 1841 but not 1834) minister of Small Isles. After the Disruption, Swanson famously served his (now Free Church) parish from a floating manse, the yacht *Betsey*, as immortalised in Miller's *Cruise of the Betsey* (2020 [1858]). The fish seemingly went to King's College, Aberdeen (Andrews 1982a, p. 67); **cave … night**: an episode vividly described in Miller 1854a, pp. 74–79; 1993, pp. 73–79.

CHAPTER VII.

Further Discoveries of the Ichthyolite Beds.—Found in one
Locality under a Bed of Peat.—Discovered in another
beneath an ancient Burying-ground.—In a third underlying the Lias formation.—In a fourth overtopped by a still
older Sandstone Deposit.—Difficulties in ascertaining the
true Place of a newly-discovered Formation.—Caution
against drawing too hasty Inferences from the mere circumstance of Neighbourhood.—The Writer receives his first Assistance from without.—*Geological Appendix* of the Messrs
Anderson of Inverness.—Further Assistance from the Researches of Agassiz.—Suggestion.—Dr John Malcolmson.—
His extensive Discoveries in Moray.—He submits to Agassiz
a drawing of the *Pterichthys*.—Place of the Ichthyolites in
the Scale at length determined.—Two distinct Platforms of
Being in the Formation to which they belong.

I COMMENCED forming a small collection, and set
myself carefully to examine the neighbouring rocks for
organisms of a similar character. The eye becomes
practised in such researches, and my labours were
soon repaid. Directly above the little bay there is
a corn-field, and beyond the field a wood of forest
trees. And in this wood, in the bottom of a water-
course, scooped out of the rock through a bed of peat,
I found the stratified clay charged with scales. A
few hundred yards farther to the west there is a deep
wooded ravine cut through a thick bed of red diluvial

clay. The top of the bank directly above is occupied
by the ruins of an ancient chapel, and a group of moss-
grown tombstones. And in the gorge of this ravine,
underlying the little field of graves by about sixty feet,
I discovered a still more ancient place of sepulture—
that of the ichthyolites. I explored every bank, rock,
and ravine on the northern or Cromarty-Frith side of
the tongue of land, with its terminal point of granitic
gneiss, to which I have had such frequent occasion
to refer, and then turned to explore the southern or
Moray-Frith side, in the rectilinear line of the great
valley. And here I was successful on a larger scale.
A range of lofty sandstone cliffs, hollowed by the
sea, extends for a distance of about two miles between two of the granitic knobs or wedges of the line,
—the Southern Sutor and the hill of Eathie. And
along well nigh the entire length of this range of cliffs
I succeeded in tracing a continuous ichthyolite bed,
abounding in remains, and lying far below the Lias,
and unconformable to it. I pursued my researches,
and in the sides of a romantic precipitous dell, through
which the Burn of Eathie—a small mossy stream—
finds its way to the Moray Frith, I again discovered
the fish-beds running deep into the interior of the
country, with immense strata of a pale yellow sandstone resting over them, and strata of a chocolate-
red lying below. But their place in the geological
scale was still to fix.

I had seen enough to convince me that they form
a continuous convex stratum in the sandstone spear-
shaft, covering it saddlewise from side to side, dipping towards the Moray Frith on the south, and to

---

chapel: of St Regulus, north-east of Cromarty House; **lying far below the Lias, and
unconformable to it**: 'far below', and with a large space of time between, in the sense
of the theoretical stratigraphic column. At Eathie, of course, the faulting brings up the
Old Red Sandstone to lie adjacent to the Jurassic beds; **country**: Black Isle, not Scotland as a whole.

the Cromarty Frith on the north,—that, as in a *bona
fide* spear-shaft, the annual ring or layer of growth
of one season is overlaid by the annual rings of succeeding seasons, and underlaid by those of preceding
ones ; so this huge semi-ring of fossiliferous clays and
limestones had its underlying semi-ring of Red Sandstone, and its overlying semi-rings of yellow, of red,
and of gray sandstone. I knew, besides, that beneath
there was a semi-ring of conglomerate, the base of
the system ; and that for more than two hundred
yards upwards, ring followed ring in unbroken succession,—now sandstone, now limestone, now stratified clay. But though intimately acquainted with
these lower rocks for more than a hundred fathoms
from their base upwards, and with the upper rocks on
both sides the ichthyolite bed for more than a hundred feet, there was an intervening hiatus, whose extent at this period I found it impossible to ascertain.
And hence my uncertainty regarding the place of the
ichthyolites, seeing that whole formations might be
represented by the occurring gap. On the Moray-
Frith side, where the sections are of huge extent, a
doubtful repeat in the strata at one point of junction,
and an abrupt fault at another, cuts off the upper
series of beds to which the organisms belong, from the
lower to which the great conglomerate belongs. On
the Cromarty-Firth side the sections are mere detached patches, obscured at every point by diluvium and
soil ; and, in conceiving of the whole as a continuous
line, with the Lias a-top and the granitic group at the
bottom, I was ever reminded of those coast-lines of
the ancient geographers, where a few uncertain dots,

a few deeper markings, and here and there a blank space or two, showed the blended results of conjecture and discovery,—whether they gave a *Terra Incognita Australia* to the one hemisphere, or a North-Western passage to the other. The ichthyolites in a section so doubtful might be regarded as belonging to either the Old or the New Red Sandstone,—to the Coal Measures or to the Mountain Limestone. All was uncertainty.

One remark in the passing : it may teach the young geologist to be cautious in his inferences, and illustrate, besides, those gaps which occur in the geological scale. I had now discovered the ichthyolite beds in five different localities ; in one of these—the first discovered—there is no overlying stratum ; it seems as if the bed formed the top of the formation : in all the others the overlying stratum is different, and belongs to distant and widely-separated ages. We cut in one locality through a peat-moss,—part of the ruins, perhaps, of one of those forests which covered, about the commencement of the Christian era, well nigh the entire surface of the island, and sheltered the naked inhabitants from the legions of Agricola. We find, as we dig, huge trunks of oak and elm, cones of the Scotch fir, handfuls of hazel-nuts, and bones and horns of the roe and the red-deer. The writer, when a boy, found among the peat the horn of a gigantic elk. And, forming the bottom of this recent deposit, and *lying conformably to it*, we find the ichthyolite beds, with their antique organisms. The remains of oak and elm leaves, and of the spikes and cones of the pine, lie within half a foot of the remains of the *Coccosteus* and *Diplopterus*. We dig in another

---

Terra Incognita Australia: more correctly *australis*; not Australia but the mythical Great Southern Continent about which there was much speculation until the eighteenth-century voyages of Cook and other explorers; **North-Western Passage**: the then-mythical ice-free sea-passage to Asia around the north of America; **peat-moss**: probably on the slope backing the flat raised beach behind the fish site (p. 118); **Agricola**: Roman governor of Britannia who led an invasion of Scotland; **elk**: by implication a Giant Deer or 'Irish Elk', but sadly this cannot be confirmed as the finds were sent to a London collector and lost sight of: Miller 1854a, pp. 67–68; 1993, pp. 67–68.

locality through an ancient burying-ground ; we pass through a superior stratum of skulls and coffins, and an inferior stratum barren in organic remains, and then arrive at the stratified clays, with their ichthyolites. In a third locality we find these in junction with the Lias, and underlying its lignites, ammonites, and belemnites, just as we find them underlying, in the other two, the human bones and the peat-moss. And in yet a fourth locality we see them overlaid by immense arenaceous beds, that belong evidently, as their mineralogical character testifies, to either the Old or the New Red Sandstone. The convulsions and revolutions of the geological world, like those of the political, are sad confounders of place and station, and bring into close fellowship the high and the low ; nor is it safe in either world,—such have been the effects of the disturbing agencies,—to judge of ancient relations by existing neighbourhoods, or of original situations by present places of occupancy. "Misery," says Shakspeare, " makes strange bed-fellows :" the changes and convulsions of the geological world have made strange bed-fellows too. I have seen fossils of the Upper Lias and of the Lower Old Red Sandstone washed together by the same wave, out of what might be taken, on a cursory survey, for the same bed, and then mingled with recent shells, algæ, branches of trees, and fragments of wrecks on the same sea-beach.

Years passed, and in 1834 I received my first assistance from without, through the kindness of the Messrs Anderson of Inverness, who this year published their *Guide to the Highlands and Islands of*

burying-ground: St Regulus's (p. 118); it was not Miller, but the Old Chapel Burn, that did the digging, and the barren stratum was in fact boulder clay, all later discussed in *Cruise of the Betsey* (Miller 2022 [1858], pp. 304–22); **third … fourth locality**: Eathie shore, and Burn of Eathie (p. 119); **'Misery … makes strange bed-fellows'**: adapted or remembered from *The Tempest*, Act II, Scene 2, line 42, 'Misery acquaints a man with strange bedfellows'; **sea-beach**: Eathie; **guide**: Anderson and Anderson 1834.

Scotland,—a work which has never received half its due measure of praise. It contains, in a condensed and very pleasing form, the accumulated gleanings for half a lifetime of two very superior men, both skilled in science, and of highly cultivated taste and literary ability. As they had repeatedly travelled over almost every foot-breadth of country which they describe, their remarks exhibit that freshness of actual observation, recorded on the spot, which Gray regarded as " worth whole cart-loads of recollection." But what chiefly interested me in their work was its dissertative appendices,—admirable digests of the Natural History, Antiquities, and Geology of the country. The appendix devoted to Geology, consisting of fifty closely-printed pages,—abridged in part from the highest geological authorities, and in much greater part the result of original observation,— contains beyond comparison the completest description of the rocks, fossils, and formations of the Northern and Western Highlands, which has yet been given to the public in a popular form. I perused it with intense interest, and learned from it, for the first time, of the fossil fishes of Caithness and Gamrie.

There was almost nothing known at the period, of the oryctology of the older rocks,—little, indeed, of that of the Old Red Sandstone, in its proper character as such ; and, with no such guiding clue as has since been furnished by Agassiz, and the later researches of Mr Murchison, the writer of the appendix had recorded as his ultimate conclusion, that " the middle schistoze system of Caithness, containing the fossil fish, was intermediate in geological character

---

two very superior men: see p. 125; **'cart-loads'**: remembered from a letter of 1753 by the English poet Thomas Gray (1716–71), 'half a word fixed upon or near the spot, is worth a cart-load of recollection' (Gray 1820, vol. II, p. 54); **appendix**: Anderson 1834, pp. 198, 205; **schistoze**: here in the older sense of laminated or layered rocks, rather than pertaining to the metamorphic rock schist. The quotation is modified from Anderson 1834, p. 198, and ends with 'New Red Sandstone formations' on p. 124.

and position between the Old and New Red Sandstone formations. The ichthyolites of Gamrie he described as resembling those of Caithness; and I at
once recognized, in his minute descriptions of both,
the fossil fish of Cromarty. The mineralogical ac-
companiments, too, seemed nearly the same. In
Caithness the animal remains are mixed up in some
places with a black bituminous matter like tar. I
had but lately found among the beds of the little bay
a mass of soft adhesive bitumen, hermetically sealed
up in the limestone, which, when broken open, reminded me, from the powerful odour it cast, and
which filled for several days the room in which I
kept it, of the old Gaulish mummy of which we find
so minute account in the Natural History of Goldsmith. The nodules which enclosed the organisms at
Gamrie were described as of a sub-crystalline, radiating, fibrous structure. So much was this the case
with some of the nodules at Cromarty, that they
had often reminded me, when freshly broken, though
composed of pure carbonate of lime, of masses of as-
bestos. The scales and bones of the Caithness ichthyolites were blended, it was stated, with the fragments
of a "supposed tortoise nearly allied to trionyx;"
one of the ichthyolites, a *Dipterus*, was characterized by large scales, a double dorsal, and a one-sided
tail; the entire lack of shells and zoophytes was remarked, and the abundance of obscure vegetable impressions. In short, had the accomplished writer of
the appendix been briefly describing the beds at
Cromarty, instead of those of Caithness and Gamrie, he might have employed the same terms, and re-

---

little bay: east of Cromarty burgh; mummy ... Goldsmith: an Auvergne mummy is
described in chapter 12 ('Of Mummies') of the 'History of Animals' in the popular
natural-history compendium *An History of the Earth and Animated Nature* (1774) by the
Irish writer Oliver Goldsmith (?1730–74); 'tortoise': from Anderson 1834, p. 198, adding
'supposed'. *Trionyx*, a freshwater terrapin, was the initial identification of the puzzling
fishes such as *Coccosteus* and *Pterichthys*, e.g. by Sedgwick and Murchison (1829b,
p. 198), who took it as evidence for the freshwater origin of the Old Red Sandstone: Andrews 1982a, p. 10; double dorsal: two dorsal fins (see p. 80).

marked the same circumstances,—the striated nodules, the mineral tar, the vegetable impressions, the
absence of shells and zoophites, the large-scaled and
double-finned ichthyolites,—the peculiarities of which
applied equally to the *Dipterus* and *Diplopterus*,—and
the supposed tortoise, in which I once recognized the
Coccosteus. It was much to know that this doubtful
formation, for as doubtful I still regarded it, was of
such considerable extent, and occurred in localities so
widely separated. I corresponded with the courteous
author of the appendix, at that time General Secretary to the Northern Institution for the Promotion of
Science and Literature, and Conservator of its Museum ; and, forwarding to him duplicates of some of
my better specimens, had, as I had anticipated, the
generic identity of the Cromarty ichthyolites with
those of Caithness and Gamrie fully confirmed.

My narrative is, I am afraid, becoming tedious ;
but it embodies somewhat more than the mere history
of a sort of Robinson Crusoe in Geology, cut off for
years from all intercourse with his kind. It contains
also the history of a formation in its connection with
science ; and the reader will, I trust, bear with me for
a few pages more. Seasons passed ; and I received
new light from the researches of Agassiz, which, if it
did not show me my way more clearly, rendered it
at least more interesting, by associating with it one
of those wonderful truths, stranger than fictions, which
rise ever and anon from the profounder depths of
science, and whose use, in their connection with the
human intellect, seems to be to stimulate the faculties. I have had often occasion to refer to the one-

---

General Secretary to the Northern Institution: George Anderson (1802–78), a keen
natural historian and antiquary, and founder of this philosophical society in 1825
(Andrews 1982a, p. 57; Wallace 1921). George and his brother Peter were Inverness
solicitors as well as authors of the *Guide to the Highlands and Islands of Scotland*:
Anderson and Anderson 1834.

sided condition of tail characteristic of the ichthyolites of the Old Red Sandstone. It characterizes, says Agassiz, the fish of all the more ancient formations. At one certain point in the descending scale nature entirely alters her plan in the formation of the tail. All the ichthyolites above are fashioned after one particular type,—all below after another and different type. The bibliographer can tell at what periods in the history of letters one character ceased to be employed and another came into use. Black letter, for instance, in our own country, was scarce ever resorted to for purposes of general literature after the reign of James VI. ; and in manuscript writing the Italian hand superseded the Saxon about the close of the seventeenth century. Now, is it not truly wonderful to find an analogous change of character in that pictorial history of the past which Geology furnishes ? From the first appearance of vertebrated existences to the middle beds of the New Red Sandstone,—a space including the Upper Ludlow rocks, the Old Red Sandstone in all its members, the Mountain Limestone with the Limestone of Burdie House, the Coal Measures, the Lower New Red, and the Magnesian Limestone,—we find only the ancient or unequally-lobed type of tail. In all the formations above, including the Lias, the Oolite, Middle, Upper, and Lower, the Wealden, the Green-sand, the Chalk, and the Tertiary, we find only the equally-lobed condition of tail. And it is more than probable, that with the tail the character of the skeleton also changed ; that the more ancient type characterized, throughout, the semi-cartilaginous order of fishes, just as the

---

Agassiz: probably from Agassiz 1834, reinforced by some press coverage of the Glasgow BAAS meeting, e.g. end matter p. 2, and Anon. 1840m. See note for p. 243.

more modern type characterizes the osseous fishes ;
and that the upper line of the Magnesian Limestone
marks the period at which the order became extinct.
Conjecture lacks footing in grappling with a revolution so extensive and so wonderful.   Shall I venture
to throw out a suggestion on the subject, in connection
with another suggestion which has emanated from
one of the first of living geologists ?   Fish, of all existing creatures, seem the most capable of sustaining
high degrees of heat, and are to be found in some of
the hot springs of Continental Europe, where it is
supposed scarce any other animal could live.   Now,
all the fish of the ancient type are thickly covered by
a defensive armour of bone, arranged in plates, bars,
or scales, or all the three modes together, as in the
Osteolepis and one half its cotemporaries.   The one-
sided tail is united invariably to a strong cuirass.   And
it has been suggested by Dr Buckland, that this strong
cuirass may have formed a sort of defence against the
injurious effects of a highly heated surrounding medium.   The suggestion is, of course, based purely on
hypothesis.   It may be stated in direct connection
with it, however, that in the Lias,—the first richly
fossiliferous formation overlying that in which the
change occurred,—we find, for the first time in the
geological system, indications of a change of seasons.
The foot-prints of winter are left impressed amid the
lignites of the Cromarty Lias.   In a specimen now
before me, the alternations of summer-heat and winter-
cold are as distinctly marked in the annual rings, as
in the pines or larches of our present forests; whereas
in the earlier lignites, cotemporary with ichthyolites

Buckland: Buckland (1836, vol. I, pp. 282–83 n.) had applied this suggestion to all fishes
older than the Cretaceous (not just the Devonian ones), as they also had thick enamelled scales. But Miller silently alters it to a tentative means of explaining the more
heavily armoured appearance of the older fishes. Robert Dick was quick to point out
the difficulty with this explanation, namely that co-existing fish genera had armour of
greatly differing thicknesses. In a letter to Miller, he parodied the hypothesis in Dicken-
sian manner by imagining two fishes, 'Cocco' and 'Osteolep', grumbling about the heat
(Smiles 1878, pp. 106–7).

of the ancient type, no annual rings appear. *Just ere winter began to take its place among the seasons, the fish fitted for living in a highly heated medium disappeared:* they were created to inhabit a thermal ocean, and died away as it cooled down. Fish of a similar type may now inhabit the seas of Venus, or even of Jupiter, which, from its enormous bulk, though greatly more distant from the sun than our earth, may still powerfully retain the internal heat.

I still pursued my inquiries, and received a valuable auxiliary in a gentleman from India, Dr John Malcolmson of Madras,—a member of the London Geological Society, and a man of high scientific attainments and great general knowledge. Above all, I found him to possess, in a remarkable degree, that spirit of research, almost amounting to a passion, which invariably marks the superior man. He had spent month after month under the burning sun of India, amid fever-marshes and tiger-jungles, acquainting himself with the unexplored geological field which, only a few years ago, that vast continent presented, and in collecting fossils hitherto unnamed and undescribed. He had pursued his inquiries, too, along the coasts of the Red Sea, and far upwards on the banks of the Nile ; and now, in returning for a time to his own country, he had brought with him the determination of knowing it thoroughly as a man of science and a geologist. I had the pleasure of first introducing him to the ichthyolites of the Lower Old Red Sandstone, by bringing him to my first-discovered bed, and laying open, by a blow of the hammer, a beautiful *Osteolepis*. He was much interested in the

---

Venus … Jupiter: several Scottish Evangelical natural philosophers, e.g. Thomas Chalmers and David Brewster, believed that other heavenly bodies were (or would be) the abodes of intelligent beings (Crowe 1986). This longstanding view would become controversial in the 1850s, with Miller involved (Brooke 1977; Jenkins 2015). Miller here envisages other planets in *earlier* phases of their progress towards perfection; **internal heat**: see p. A63; **Malcolmson**: Dr John G. Malcolmson (1802–44), army surgeon with the East India Company and Forres native, on home leave (Keillar 1993). His visit was 'at the close of 1837' (Miller 1854a, p. 507, 1993, p. 505; Andrews 1982a, pp. 67–69).

fossils of my little collection, and at once decided that
the formation which contained them could be no representative of the Coal Measures. After ranging over
the various beds on both sides the rectilinear ridge, and
acquainting himself thoroughly with their organisms,
he set out to explore the Lower Old Red Sandstones of
Moray and Banff, hitherto deemed peculiarly barren,
but whose character too much resembled that of the
rocks which he had now ascertained to be so abundant
in fossils not to be held worthy of farther examination.
He explored the banks of the Spey, and found the ichthyolite beds extensively developed at Dipple, in the
middle of an Old Red Sandstone district. He pursued his researches, and traced the formation in ravines and the beds of rivers, from the village of
Buckie to near the field of Culloden; he found it
exposed in the banks of the Nairn, in the ravines
above Cawdor Castle, on the eastern side of the hill
of Rait, at Clune, Lethenbar, and in the vale of
Rothes,—and in every instance low in the Old Red
Sandstone. The formation hitherto deemed so barren in remains proved one of the richest of them all,
if not in tribes and families, at least in individual
fossils; and the reader may form some idea of the
extent in which it has already been proved fossiliferous, when he remembers that the tract includes as its
extremes, Orkney, Gamrie, and the north-eastern
gorge of the great Caledonian Valley. The ichthyolites were discovered in the latter locality in the quarry
of Inches, three miles beyond Inverness, by Mr George
Anderson, the gentleman to whose geological attain-

I

Dipple: on the left bank of the River Spey opposite Fochabers; **Culloden**: the 1746
battlefield, about 5 miles east of Inverness; **Cawdor Castle**: about 5 miles south-west of
Nairn; **hill of Rait**: presumably that about 4 miles south of Nairn; **Lethen Bar**: now
also Cairn Bar, a hill about 7 miles south-east of Nairn, but in this context a small
quarry on the north-eastern slopes, near that at **Clune** – both of which yielded fish-
bearing nodules, burnt for lime. See note for p. 134; **vale of Rothes**: from Malcolmson
(1842, p. 142), apparently Birnie just south of Elgin; **gorge**: throat, as in military fortifi-
cation; the north end of the Great Glen; **Inches**: or Inshes, south-east of Inverness.

ments, as one of the authors of the *Guide Book*, I have lately had occasion to refer.

I had now corresponded for several years with a little circle of geological friends, and had described in my letters, and in some instances had attempted to figure in them, my newly-found fossils. A letter which I wrote early in 1838 to Dr Malcolmson, then at Paris, and which contained a rude drawing of the *Ptericthys*, was submitted to Agassiz, and the curiosity of the naturalist was excited. He examined the figure, rather, however, with interest than surprise, and read the accompanying description, not in the least inclined to scepticism by the singularity of its details. He had looked on too many wonders of a similar cast to believe that he had exhausted them, or to evince any astonishment that Geology should be found to contain one wonder more. Some months after, I sent a restored drawing of the same fossil to the Elgin Scientific Society. I must state, however, that the restoration was by no means complete. The paddle-like arms were placed farther below the shoulders than in any actual species; and I had transferred, by mistake, to the creature's upper side some of the plates of the *Coccosteus*. Still the type was unequivocally that of the *Ptericthys*. The Secretary of the Society, Mr Patrick Duff, an excellent geologist, to whose labours, in an upper formation of the Old Red Sandstone, I shall have afterwards occasion to refer, questioned, as he well might, some of the details of the figure, and we corresponded for several weeks regarding it, somewhat in the style

---

Malcolmson: see note for p. 47 and, for this episode as a whole, Andrews 1982a, pp. 23–26; **sent … to the Elgin Scientific Society**: letter to Duff on 15 December 1838, HMLB item 207 = ELGNM Geology Letter G1/2 (Figure 20); **Duff**: Patrick Duff (1791–1861), Town Clerk of Elgin, important local geologist and fossil collector, and Secretary of the Elgin and Moray Scientific Association: Andrews 1982a, pp. 63–64.

of Jonathan Oldbuck and his antiquarian friend, who succeeded in settling the meaning of two whole words, in an antique inscription, in little more than two years. Most of the other members looked upon the entire drawing, so strange did the appearance seem, as embodying a fiction of the same class with those embodied in the pictured griffins and unicorns of mythologic Zoology; and, in amusing themselves with it, they bestowed on its betailed and bepaddled figure, as if in anticipation of Agassiz, the name of the draughtsman. Not many months after, however, a true *bona fidePtericthys* turned up in one of the newly-discovered beds of Nairnshire, and the Association ceased to joke, and began to wonder. I merely mention the circumstance in connection with a right challenged, at the late meeting of the British Association at Glasgow, by a gentleman of Elgin, to be regarded as the original discoverer of the *Ptericthys*. I am of course far from supposing that the discovery was not actually made, but regret that it should have been kept so close a secret at a time when it might have stood the other discoverer of the creature in such stead.

The exact place of the ichthyolites in the system was still to fix. I was spending a day early in the winter of 1839 among the nearly vertical strata that lean against the Northern Sutor. The section there presented is washed by the tide for nearly three hundred yards from where it rests on the granitic gneiss; and each succeeding stratum in the ascending order may be as clearly traced as the alternate white and black squares in a marble pavement. First there is

---

Oldbuck and his ... friend: characters in Scott's novel *The Antiquary* (1816); **draughtsman**: Miller himself; ***bona fide Ptericthys***: found at Lethen Bar on 27 March 1839 by William Stables junior, factor for the Cawdor estate and keen naturalist: Gordon 1859, pp. 20–22; Andrews 1982a, p. 26; Noltie 2012; **gentleman of Elgin**: one William Keir or Kier, who seems to have been an itinerant lecturer rather than an inhabitant of Elgin: Gordon 1859, p. 23; Andrews 1982a, pp. 27–28, 74 n.150.

132   THE OLD RED SANDSTONE.

a bed of conglomerate two hundred and fifteen feet
in thickness, " identical in structure," say Professor Sedgwick and Mr Murchison, " with the older
red conglomerates of Cumberland and the Island of
Arran,* and which cannot be distinguished from the
conglomerates which lean against the southern flank
of the Grampians, and on which Dunnottar Castle is
built.   Immediately above the conglomerate there is
a hundred and fourteen feet more of coarse sandstone
strata, of a reddish-yellow hue, with occasionally a
few pebbles inclosed, and then twenty-seven feet additional of limestone and stratified clay.   There are no
breaks, no faults, no thinning out of strata,—all the
beds lie parallel, showing regular deposition.   I had
passed over the section twenty times before, and had
carefully examined the limestone and the clay, but in
vain.   On this occasion, however, I was more fortu-
nate.   I struck off a fragment.   It contained a vegetable impression of the same character with those of
the ichthyolite beds ; and after an hour's diligent
search, I had turned from out the heart of the stratum,
plates and scales enough to fill a shelf in a museum,
—the helmet-like snout of a *Diplopterus*, the thorn-
like spine of a *Cheiracanthus*, and a *Coccosteus* well
nigh entire.   I had at length, after a search of nearly
ten years, found the true place of the ichthyolite

* Different in one respect from the conglomerates of Arran.
It abounds in rolled fragments of granite, whereas in those
of Arran there occur no pebbles of this rock.   Arran has now
its granite in abundance ; the northern locality has none ;
though, when the conglomerates of the Lower Old Red Sandstone were in the course of forming, the case was exactly the
reverse.

'identical …': Sedgwick and Murchison 1829b, p.156, much edited down (the cited
passage lacks closing quotation marks but in fact ends with **Grampians**). They were
speaking of the 'old conglomerates and sandstones of Caithness and the M[oray] Firth'
rather than the North Sutor conglomerate alone; **Arran**: probably from Sedgwick and
Murchison 1829a, pp. 25, 34 and n.

bed. The reader may smile, but I hope the smile will be a good-natured one : a simple pleasure may be not the less sincere on account of its simplicity; and " little things are great to little men." I passed over and over the strata, and found there could be no mistake. The place of the fossil fish in the scale is little more than a hundred feet above the top, and not much more than a hundred yards above the base of the great conglomerate : and there lie over it in this section about five hundred feet of soft arenaceous stone, with here and there alternating bands of lime-stone and beds of clay studded with nodules,—all be-longing to the inferior Old Red Sandstone.

The enormous depth of the Old Red Sandstone of England has been divided by Mr Murchison into three members or formations,—the division adopted in his *Elements* by Mr Lyall, as quoted in an early chapter. These are, the lowest or Tilestone formation, the middle or Cornstone formation, and the uppermost or Quartzose-conglomerate formation. The terms are derived from mineralogical character, and inadequate as designations, therefore, like that of the Old Red Sandstone itself, which in many of its deposits is not *sandstone*, and is not *red*. But they serve to express great natural divisions. Now the Tilestone member of England represents, as I have already stated, this Lower Old Red Sandstone formation of Scotland ; but its extent of vertical development, compared with that of the other two members of the system, is strik-ingly different in the two countries. The Tilestones compose the least of the three divisions in England ;— their representative in Scotland forms by much the

---

'little things are great to little men': a proverbial expression in general use in the early nineteenth century, a variant of the earlier proverb 'little things please little minds'. He had reason to be pleased, for his find was almost certainly in the 'few weeks' which he 'gave … very sedulously to geology' between resigning from the bank at the end of its financial year, and leaving for Edinburgh (Miller 1854a, pp. 535–36, 1994, p. 534); **Murchison**: in Murchison 1839, part I, pp. 170–71; **Lyall**: error for Lyell.

greatest of the three ; and there seems to be zoological as well as lithological evidence that its formation must have occupied no brief period. *The same genera occur in its upper as in its lower beds, but all the species appear to be different.* I shall briefly state the evidence of this very curious fact.

The seat of Sir William Gordon Cumming of Altyre is in the neighbourhood of one of the Morayshire deposits discovered by Dr Malcolmson ; and for the greater part of the last two years Lady Gordon Cumming has been engaged in making a collection of its peculiar fossils, which already fills an entire apartment. The object of her Ladyship was the illustration of the Geology of the district, and all she sought in it on her own behalf was congenial employment for a singularly elegant and comprehensive mind. But her labours have rendered her a benefactor to science. Her collection was visited, shortly after the late meeting of the British Association in Glasgow, by Agassiz and Dr Buckland; and great was the surprise and delight of the philosophers to find that the whole was new to Geology. All the species, amounting to eleven, and at least one of the genera, that of the *Glyptolepis*, were different from any Agassiz had ever seen or described before. The deposit so successfully explored by her Ladyship occurs high in the lower formation. Agassiz, shortly after, in comparing the collection of Dr Traill—a collection formed at Orkney,—with that of the writer—a collection made at Cromarty,—was struck by the specific identity of the specimens. In the instances in which the genera agreed he found that the species agreed also,

---

one of the ... deposits: Lady Gordon Cumming's collection (see p. 36) came in large part from the hill of Lethen Bar (see p. 129), and perhaps also from the adjacent site of Clune which worked the same fish bed. There has been considerable confusion on this matter over the years: Malcolmson 1842; Miller 2022 [1858], p. 192; Andrews 1982a, pp. 71 n.115, 72–73 n.123; Andrews 1983; visited ... by Agassiz: October 1840; see Appendix 2.

though the ichthyolites of both differed specifically from the ichthyolites of Caithness, which occur chiefly in the middle and upper beds of the formation, and from those also of Lady Cumming of Altyre, which occur, as I have said, at the top. And in examining into the cause, it was found that the two collections, though furnished by localities more than a hundred miles apart, were yet derived, if I may so express myself, from the same low platform, both alike representing the fossiliferous base of the system, and both removed but by a single stage from the great unfossiliferous conglomerate below. Thus there seem to be what may be termed two storeys of being in this lower formation—storeys in which the groupes, though generically identical, are specifically dissimilar.

---

platform: not, of course, a topographical or physical platform, but Miller's word for what one would normally call a stratum or horizon; formation: i.e. his Lower Old Red Sandstone; generically and specifically: referring to the genus and species of Linnean taxonomy.

CHAPTER VIII.

Upper Formations of the Old Red Sandstone.—Room enough for each and to spare.—Middle or Cornstone Formation.—The *Cephalaspis* its most characteristic Organism.—Description.—The Den of Balruddery richer in the Fossils of this middle Formation than any other Locality yet discovered.—Various Cotemporaries of the *Cephalaspis.*—Vegetable Impressions.—Gigantic Crustacean.—*Seraphim.*—Ichthyodorulites.—The Upper Formation.—Wide Extent of the Fauna and Flora of the earlier Formations.—Probable Cause.

HITHERTO I have dwelt almost exclusively on the fossils of the Lower Old Red Sandstone, and the history of their discovery : I shall now ascend to the organisms of its higher platforms. The system in Scotland, as in the sister kindom, has its middle and upper groupes, and these are in no degree less curious than the inferior group already described, nor do they less resemble the existences of the present time. Does the reader remember the illustration of the pyramid employed in an early chapter,—its three parallel bars, and the strange hieroglyphics of the middle bar ? Let him now imagine another pyramid, inscribed with the remaining and later history of the system. We read as before from the base upwards, but find the broken and half-defaced characters of the second erection

---

sister kindom: emended in 2nd edition to 'kingdom', i.e. England; **pyramid ... earlier chapter:** pp. 30–33.

descending into the very soil, as in those obelisks of
Egypt round which the sands of the desert have been
accumulating for ages. Hence a hiatus in our history, for future excavators to fill; and it contains many
such blanks, every unfossiliferous bar in either pyramid representing a gap in the record. Three distinct formations the group undoubtedly contains,—
perhaps more; nor will the fact appear strange to the
reader who remembers how numerous the formations
are that lie over and under it, and that its vast depth
of ten thousand feet equals that of the whole secondary system from top to bottom. Eight such formations as the Oolite, or ten such formations as the
Chalk, could rest, the one over the other, in the space
occupied by a group so enormous. To the evidence
of its three distinct formations, which is of a very
simple character, I shall advert as I go along.

The central or Cornstone division of the system in
England is characterized throughout its vast depth by
a peculiar family of ichthyolites, which occur in none
of the other divisions. I have already had occasion
to refer to the *Cephalaspis*. Four species of this fish
have been discovered in the Cornstones of Hereford,
Salop, Worcester, Monmouth, and Brecon; " and as
they are always found," says Mr Murchison, " in the
same division of the Old Red System, they have become valuable auxiliaries in enabling the geologist to
identify its subdivisions through England and Wales,
and also to institute direct comparisons between the
different strata of the Old Red Sandstone of England
and Scotland." The *Cephalaspis* is one of the most
curious ichthyolites of the system. (See Plate IX.

---

group: i.e. the Old Red Sandstone system as a whole, *not* the 'pyramid'; **Four species
... Murchison**: Murchison 1839, part II, p. 587; Agassiz 1833–43 and in Murchison 1839,
part II, pp. 589–96.

fig. 1.) Has the reader ever seen a saddler's cutting knife?—a tool with a crescent-shaped blade, and the handle fixed transversely in the centre of its concave side. In general outline the *Cephalaspis* resembled this tool,—the crescent-shaped blade representing the head,—the transverse handle the body. We have but to give the handle an angular instead of a rounded shape, and to press together the pointed horns of the crescent till they incline towards each other, and the convex or sharpened edge is elongated into a semi-ellipse, cut in the line of its shortest diameter, in order to produce the complete form of the *Cephalaspis*. The head, compared with the body, was of great size, —comprising fully one-third the creature's entire length. In the centre, and placed closely together, as in many of the flat fish, were the eyes. Some of the specimens show two dorsals, and an anal and caudal fin. The thin and angular body presents a jointed appearance, somewhat like that of a lobster or trilobite. Like the bodies of most of the ichthyolites of the system, it was covered with variously-formed scales of bone; the creature's head was cased in strong plates of the same material, the whole upper side lying under one huge buckler,—and hence the name *Cephalaspis*, or buckler-head. In proportion to its strength and size, it seems to have been amply furnished with weapons of defence. Such was the strength and massiveness of its covering, that its remains are found comparatively entire in arenaceous rocks impregnated with iron, in which few other fossils could have survived. Its various species, as they occur in the Welsh and English Cornstones, says Mr

---

iron: see p.156.

Murchison, seem " not to have been suddenly killed and entombed, but to have been long exposed to submarine agencies, such as the attacks of animals, currents, concretionary action," &c. ; and yet, " though much dismembered, the geologist has little difficulty in recognising even the smallest portions of them." Nor does it seem to have been quite unfurnished with offensive weapons. The sword-fish, with its strong and pointed spear, has been known to perforate the oaken ribs of the firmest-built vessels ; and, poised and directed by its lesser fins, and impelled by its powerful tail, it may be regarded either as an arrow or javelin flung with tremendous force, or as a knight speeding to the encounter with his lance in rest. Now there are missiles employed in eastern warfare, which, instead of being pointed like the arrow or javelin, are edged somewhat like the crooked falchion or saddler's cutting-knife, and which are capable of being cast with such force, that they have been known to sever a horse's leg through the bone; and if the sword-fish may be properly compared to an arrow or javelin, the combative powers of the *Cephalaspis* may be illustrated, it is probable, by a weapon of this kind,—the head all around its elliptical margin presenting a sharp edge, like that of a cutting-knife or falchion. Its impetus, however, must have been comparatively small, for its organs of motion were so : it was a bolt carefully fashioned, but a bolt cast from a feeble bow. But if weak in the assault, it must have been formidable when assailed. " The pointed horns of the crescent," said Agassiz to the writer, " seem to have served a similar purpose with the spear-like wings of the *Pter-*

Murchison: paraphrased from Murchison 1839, part II, p. 588.

icthys,"—the sole difference consisting in the circumstance, that the spears of the one could be elevated or
depressed at pleasure, whereas those of the other were
ever fixed in the warlike attitude. And such was the
Cephalaspis of the Cornstones,—not only the most characteristic, but in England and Wales almost the sole
organism of the formation.

Now of this curious ichthyolite we find no trace
among the fossils of the Lower Old Red Sandstone.
It occurs neither in Orkney nor Cromarty, Caithness
nor Gamrie, Nairnshire nor the inferior ichthyolite
beds of Moray. Neither in England nor in Scotland
is it to be found in the Tilestone formation, or its
equivalent. It abounds, however, in the Old Red
Sandstone of Forfarshire ; it has been discovered by
Dr Malcolmson in a middle formation in Moray ; and
it occurs at Balruddery, in the Gray Sandstones which
form on both sides the Tay, where the Tilestone formation seems wanting, the apparent base of the system. It is exclusively a medal of the middle empire.

In the last-mentioned locality, in a beautifully-wood-
ed dell known as the Den of Balruddery, the *Cephalaspis* is found associated with an entire group of other
fossils, the recent discovery of Mr Webster, the proprie-
tor, who, with a zeal through which geological knowledge promises to be materially extended, and at an
expense of much labour, has made a collection of all
the organisms of the Den yet discovered. These the
writer had the pleasure of examining in the company
of Mr Murchison and Dr Buckland : he was afterwards present when they were examined by Agassiz :
and not a single organism of the group could be iden-

---

discovered by Dr Malcolmson: see Appendix 3; Balruddery: about 6 miles west of
Dundee; Webster: Robert Webster (1802–81), though the sources confuse him with
his father James (or brother of that name). It is unclear whether Robert was an active
collector (as implied on p.150), or simply inherited the collection: Andrews 1982a,
p.75 n.158; Examining ... afterward: in Edinburgh, October 1840; see Appendix 2.

Fleming (p.141): Rev. Prof. John Fleming (1785–1857), Professor of Natural Philosophy,
minister of Clackmannan, previously of Flisk in Fife, and an important zoologist and

tified on either occasion, by any member of the party, with those of the lower or upper formations. Even the genera are dissimilar. The fossils of the Lias scarce differ more from those of the Coal Measures, than the fossils of the Middle Old Red Sandstone from the fossils of the formations that rest over and under them. Each formation has its distinct group, —a fact so important to the geologist, that he may feel an interest in its further verification through the decision of yet another high authority. The superior Old Red Sandstones of Scotland were first ascertained to be fossiliferous by Professor Fleming of King's College, Aberdeen,* confessedly one of the

---

* The Upper Old Red Sandstones of Moray were ascertained to be fossiliferous at nearly the same time by Mr Martin of the Anderson Institution, Elgin. There is a mouldering conglomerate precipice termed the *Scat-Craig*, about four miles to the south of the town, more abundant in remains than perhaps any of the other deposits of the formation yet discovered; and in this precipice Mr Martin first commenced his labours in the Red Sandstones of the district, and found it a mine of wonders. It is a place of singular interest,—a rock of sepulchres; and its teeth, scales, and single bones occur in a state of great entireness; though, ere the deposit was formed, the various ichthyolites whose remains it contains seem to have been broken up, and their fragments scattered. Accumulations of larger and smaller pebbles alternate in the strata; and the bulkier bones and teeth are found invariably among the bulkier pebbles, thus showing that they were operated upon by the same laws of motion which operated on the inorganic contents of the deposit. At a considerably later period the fossils of the upper group were detected in the precipitous and romantic banks of the Findhorn, by Dr Malcolmson of Madras, when prosecuting his discoveries of the organisms of the lower formation. He found them also, though in less abundance, in a splendid section exhibited in the Burn of Lethen, a rivulet

---

geologist: Andrews 1982a, pp. 57–58 n. 31; Burns 2007; Moore 2009; **Martin**: John Martin (1800–81), schoolmaster at Anderson's Free School in Elgin, and Curator of the Elgin and Moray Scientific Association: Andrews 1982a, p. 65 n.81; *Scat-Craig*: Scaat Craig; **laws of motion**: controlling the movement of sediment particles in a water current; **Findhorn**: the stretch from Sluie to Cothall, 3 miles south-west of Forres, pp. 215–19: Malcolmson 1859, pp. 341–42; Gordon 1859, pp. 31–33; Malcolmson 1921, pp. 447–52; **Burn of Lethen**: south-west of Nairn, also called Muckle or Meikle Burn: Malcolmson 1859, p. 343–44, and in Gordon 1859, pp. 33–34; Malcolmson 1921, pp. 452–56.

first naturalists of the age, and who, to his minute acquaintance with existing forms of being, adds an acquaintance scarcely less minute with those forms of primeval life that no longer exist.   He it was who first discovered, in the Upper Old Red Sandstones of Fifeshire, the large scales and plates of that strikingly characteristic ichthyolite of the higher forma-

of Moray, and yet again in the neighbourhood of Altyre. The Rev. Mr Gordon of Birnie, and Mr Robertson of Inverugie, have been also discoverers in the district. To the geological labours of Mr Patrick Duff of Elgin, in the same field, I have already had occasion incidentally to refer. The patient enquiries of this gentleman have been prosecuted for years in all the formations of the province, from the Wealden of Linksfield, with its peculiar lacustrine remains,—lignites, minute fresh-water shells, and the teeth, spines, and vertebræ of fish and saurians,—down to the base of the Old Red Sandstone, with its *Coccostei*, *Dipteri*, and *Pterichthyes*. His acquaintance with the organisms of the *Scat-Craig* is at once more extensive and minute than that of perhaps any other geologist ; and his collection of them very valuable, representing, as it does, a formation of much interest, still little known. Mr Duff is at present engaged on a volume descriptive of the Geology of the province of Moray, a district extensively explored of late years, and abundant in its distinct groupes of organisms, but of which general readers have still much to learn ; and from no one could they learn more regarding it than from Mr Duff. It is still only a few months since the Upper Old Red Sandstones of the southern districts of Scotland were found to be fossiliferous ; and the writer is chiefly indebted for his acquaintance with their organisms to a tradesman of Berwickshire, Mr William Stevenson of Dunse, who, on perusing some of the geological articles which appeared in the *Witness* newspaper, during the course of the last autumn, sent him a parcel of fossils disinterred from out the deep belt of Red Sandstone which leans, to the south in that locality, against the grauwacke of the Lammermuirs. Mr Stevenson had recently discovered them, he stated, near Preston-haugh,

first naturalists: finest; **ichthyolite**: Fleming 1831; **Altyre**: see p. 134; **Gordon**: Rev. Dr George Gordon (1801–93), minister of Birnie near Elgin, naturalist and geologist, and co-founder of the Elgin and Moray Scientific Association: Andrews 1982a, pp. 62–63 n. 77; Keillar and Smith 1995; Bennett 2010; **Robertson**: Alexander Robertson (1816–53) of Inverugie, near Burghead, Moray; geologist, and another Association member: Andrews 1982a, pp. 66–67 n. 88; **Linksfield**: a site near Elgin: Malcolmson 1838; Miller 2022 [1858], pp. 199–202, 280–82 and B23; **volume**: Duff 1842; **Stevenson**: William Stevenson (1820–83): Duns 1885; Andrews 1982a, p. 74 n. 144.

tion, now known as the *Holyptychius*,—of which more anon; and, unquestionably, no one acquainted with his writings, or the character of his mind, can doubt that he examined carefully. Now, a few years since, I had the pleasure of introducing Professor Fleming to the organisms of the Lower Old Red Sandstone, as they occur in the neighbourhood of Cromarty; and notwithstanding his extensive acquaintance with the upper fossils of the system, he found himself, among the lower, in an entirely new field. His knowledge of the one group served but to show him how very different it was from the other. With the organisms of the lower he minutely acquainted himself: he collected specimens from Gamrie, Caithness, and Cromarty, and studied their peculiarities; and yet, on being introduced last year to the discoveries of Mr Webster at Balruddery, he found his acquaintance with both the upper and lower groupes stand him in but the same stead that his first acquired knowledge of the upper group had stood him a few years before. He agreed with Agassiz in pronouncing the group at Balruddery essentially a new group. Add to this evidence the well-weighed testimony of Mr Murchison regarding the three formations which the Old Red Sandstone contains in England, where the entire system is found continuous, the

about two miles north of Dunse, in a fine section of alternating Sandstone and conglomerate strata that lie unconformably on the grauwacke. They consisted of scales and occipital plates of the *Holoptychius*, with the remains of a bulky but very imperfectly-preserved ichthyodorulite; and the coarse arenaceous matrices which surrounded them seemed identical with the red gritty Sandstones of the Findhorn and the *Scat-Craig*.

a few years since: presumably at the end of 1837 or early to mid-1838, as it was after Malcolmson's visit 'at the close of 1837' (p. 128; Miller 1854a, p. 507), and as the sites listed do not include those discovered by Malcolmson in Moray and Nairnshire in autumn 1838 onwards; last year: presumably at Edinburgh in October 1840 (Appendix 2).

Cornstone overlying the Tilestone, and the Quartzose conglomerate the Cornstone,—take into account the fact that there, each formation has its characteristic fossil, identical with some characteristic fossil of the corresponding formation of Scotland,—that the Tilestones of the one, and the lower group of the other, have their *Dipterus* in common,—that the Cornstones of the one, and the middle group of the other, have their *Cephalaspis* in common,—that the Quartzose conglomerate of the one, and the upper group of the other, have their *Holoptychius* in common,—and then say whether the proofs of distinct succeeding formations can be more surely established. If, however, the reader still entertain a doubt, let him consult the singularly instructive section of the entire system, from the Carboniferous Limestone to the Upper Silurian, given by Mr Murchison in his *Silurian System* (Part II. Plate XXXI. fig. 1), and he will find the doubt vanish. But to return to the fossils of the Cornstone group.

In only one instance have these yet been found in Scotland north of the Grampians, and to that one I have already had occasion to refer. A species of *Cephalaspis* has been discovered in the Middle Sandstones of Moray, by Dr Malcolmson. If well sought for, they would be also found, I doubt not, in Easter-Ross in the neighbourhood of Cadboll, where there occurs, high over the ichthyolite beds, a dark gray sandstone, identical in mineral character with the gray sandstones of Balruddery. The characteristic fossil of the group, the *Cephalaspis*, occurs in considerable abundance in Forfarshire, and in a much more entire state than in the

---

section: see also sections in Miller's frontispiece, where the respective source-strata of the fossil fishes in question are indicated.

Cornstones of England and Wales. The rocks to which it belongs are also developed, though more sparingly, in the northern extremity of Fife, in a line parallel to the southern shores of the Tay. But of all the localities yet known, the Den of Balruddery is that in which the peculiar organisms of the formation may be studied with best effect. The oryctology of the Cornstones of England seems restricted to four species of the *Cephalaspis*. In Fife all the organisms of the formation yet discovered are exclusively vegetable,—darkened impressions of stems like those of the inferior ichthyolite beds, confusedly mixed with what seem slender and pointed leaflets drawn in black, and numerous circular forms which have been deemed the remains of the seed-vessels of some unknown subaerial plant. " These last occur," says Professor Fleming, the original discoverer, " in the form of circular flat patches, not equalling an inch in diameter, and composed of numerous smaller contiguous circular pieces,"—the *tout ensemble* resembling " what might be expected to result from a compressed berry, such as the bramble or the rasp." In Forfarshire the remains of the *Cephalaspis* are found associated with impressions of a different character, though equally obscure,—impressions of polished surfaces carved into seeming scales ; but in Balruddery alone are the vegetable impressions of the one locality, and the scaly impressions of the other, together with the characteristic ichthyolites of England and Forfarshire, found associated with numerous fossils besides, many of them obscure, but all of them of interest, and all of them new to Geology.

K

---

four species of *Cephalaspis*: see note for p. 137; **Fleming**: Fleming 1831, p. 86. The fossil was later named *Parka decipiens*; see note for p. 150; **rasp**: raspberry.

One of the strangest organisms of the formation is
a fossil lobster, of such huge proportions, that one of
the average-sized lobsters, common in our markets,
might stretch its entire length across the continuous
tail-flap in which the creature terminated.   And it
is a marked characteristic of the fossil, that the terminal flap should be continuous : in all the existing
varieties with which I am acquainted it is divided into
angular sections. The claws nearly resembled those of
the common lobster : their outline is similar; there
is the same hawk-bill curvature outside, and the inner
sides of the pincers are armed with similar teeth-like
tubercules.   The immense shield which covered the
upper part of the creature's body is more angular than
in the existing varieties, and resembles, both in form
and size, one of those lozenge-shaped shields worn by
knights of the middle ages on gala days, rather for
ornament than use, and on which the herald still in-
scribes the armorial bearings of ladies who bear title
in their own right.   As shown in some of the larger
specimens, the length of this gigantic crustaceon must
have exceeded four feet.   Its shelly armour was delicately fretted with the forms of circular or elliptical
scales.   On all the many plates of which it was composed we see these described by gracefully-waved
lines, and rising apparently from under one another,
row beyond row.   They were, however, as much the
mere semblance of scales as those relieved by the
sculptor on the corslet of a warrior's effigy on a Gothic
tomb,—mere sculpturings on the surface of the shell.
This peculiarity may be regarded as throwing light on
the hitherto doubtful impressions of the sandstone of

---

fossil lobster: in the generic sense of a large crustacean; **existing varieties**: living species.

Forfarshire,—impressions, as has. been said, of smooth
surfaces carved into seeming scales. They occur as
impressions merely, the sandstone retaining no more
of the original substance of the organism than the impressed wax does of the substance of the seal; and the
workmen in the quarries in which they occur, finding
form without body, and struck by the resemblance
which the delicately-waved scales bear to the sculptured markings on the wings of cherubs,—of all subjects of the chisel the most common,—fancifully
termed them *Seraphim*. They will turn out, it is
probable, to be the detached plates of some such crustaceon as the lobster of Balruddery.

The ability displayed by Cuvier in restoring from
a few broken fragments of bone the skeleton of the
entire animal to which the fragments had belonged,
astonished the world. He had learned to interpret
signs as incomprehensible to every one else as the
mysterious handwriting on the wall had been to the
courtiers of Belshazzar. The condyle of a jaw became in his hands a key to the character of the original possessor; and in a few mouldering vertebræ, or
in the dilapidated bones of a fore arm or a foot, he
could read a curious history of habits and instincts.
In common with several gentlemen of Edinburgh, all
men known to science, I was as much struck with
the skill displayed by Agassiz in piecing together
the fragments of the huge crustaceon of Balruddery,
and in demonstrating its nature as such. The numerous specimens of Mr Webster were opened out before
us. On a previous morning I had examined them,
as I have said, in the company of Mr Murchison and

---

cherubs: an order of angel in Hebrew tradition (Hebrew plural: cherubim); the comparison is to their wings as carved on gravestones; *Seraphim*: another order of angel
(singular: seraph); **They will turn out**: emended in 2nd edition to 'They have turned
out' after new evidence emerged; new text and a new plate illustrating *Seraphim* were
added (Fig. 56); **handwriting on the wall**: supernatural writing that appeared during a
feast of King Belshazzar of Babylon to prophesy the fall of his empire (Bible, Daniel 5);
crustaceon: should read 'crustacean' (author's erratum, emended in 2nd ed.); **examined**: see Appendix 2.

Dr Buckland; they had been seen also by Lord Greenock, Dr Traill, and Mr Charles M'Laren; and their fragments of new and undescribed fishes had been at once recognized with reference to at least their class. But the collection contained organisms of a different kind, which seemed inexplicable to all,—forms of various design, but so regularly mathematical in their outlines that they might be all described by a ruler and a pair of compasses, and yet the whole were covered by seeming scales. There were the fragments of scaly rhombs, of scaly crescents, of scaly circles with scaly parallelograms attached to them, and of several other regular compound figures besides. Mr Murchison, familiar with the older fossils, remarked the close resemblance of the seeming scales to those of the *Seraphim* of Forfarshire, but deferred the whole to the judgment of Agassiz; no one else hazarded a conjecture. Agassiz glanced over the collection. One specimen especially caught his attention,—an elegantly symmetrical one. It seemed a combination of the parallelogram and the crescent: there were pointed horns at each end; but the convex and concave lines of the opposite sides passed into almost parallel right lines toward the centre. His eye brightened as he contemplated it. "I will tell you," he said, turning to the company,—"I will tell you what these are,— the remains of a huge lobster." He arranged the specimens in the group before him with as much apparent ease as I have seen a young girl arranging the pieces of ivory or mother of pearl in an Indian puzzle. A few broken pieces completed the lozenge-

---

Greenock: Charles M. Cathcart (1783–1859), Lord Greenock, officer commanding the army in Scotland and Governor of Edinburgh Castle; a keen geologist whose collection of fossils from the Carboniferous of the Edinburgh area was examined by Agassiz (Andrews 1982a, p. 59 n. 41; Boase and Lunt 2004); **M'Laren**: see note for p. 18; **arranged**: Agassiz was not simply physically fitting together the broken parts of one complete individual jigsaw-wise. He was achieving something far more difficult in perceiving the relationships between isolated plates from different individuals (perhaps of different sizes and preserved in somewhat different orientations and distortions)

shaped shield ; two detached specimens placed on its opposite sides furnished the claws ; two or three semi-rings with serrated edges composed the jointed body ; the compound figure, which but a minute before had so strongly attracted his attention, furnished the ter-minal flap ; and there lay the huge lobster before us, palpable to all. There is homage due to superemi-nent genius, which nature spontaneously pays when there are no low feelings of envy or jealousy to inter-fere with her operations ; and the reader may well be-lieve that it was willingly rendered on this occasion to the genius of Agassiz.

There occur among the other organisms of Balrud-dery numerous ichthyodorulites,—fin-spines, such as those to which I have called the attention of the reader in describing the thorny-finned fish of the lower formation. But the ichthyodorulites of Bal-ruddery differ essentially from those of Caithness, Moray, and Cromarty. These last are described on both sides, in every instance, by either straight or slightly-curved lines, whereas one of the describing lines in a Balruddery variety is broken by projecting prickles that resemble sharp hooked teeth set in a jaw, or rather the entire ichthyodorulite resembles the sprig of a wild rose-bush bearing its peculiar aquiline-shaped thorns on one of its sides. Buckland in his *Bridgewater Treatise*, and Lyell in his *Ele-ments*, refer to this peculiarity of structure in ichthyo-dorulites of the latter formations. The hooks are in-variably ranged on the concave or posterior edge of the spine, and were employed, it is supposed, in ele-vating the fin. Another ichthyodorulite of the forma-

---

and from different species for all one knew, though the common surface texture would imply a single species, at least as a working hypothesis. See Figure 59.

tion resembles, in the Gothic cast of its roddings, those of the unnamed ichthyolite of the Lower Old Red Sandstone described in pages 92 and 93 of the present volume, and figured in Plate VIII. fig. 2, except that it was proportionally stouter, and traversed at its base by lines running counter to the striæ that furrow it longitudinally. Of the other organisms of Balruddery I cannot pretend to speak with any degree of certainty. Some of them seem to have belonged to the *Radiata;* some are of so doubtful a character that it can scarce be determined whether they took their place among the forms of the vegetable or animal kingdoms. One organism in particular, which was at first deemed the jointed stem of some plant resembling a calamite of the Coal Measures, was found by Agassiz to be the slender limb of a crustacean. A minute description of this interesting deposit, with illustrative prints, would be of importance to science : it would serve to fill a gap in the scale. The geological pathway, which leads upwards to the present time from those ancient formations in which organic existence first began, has been the work of well nigh as many hands as some of our longer railroads : each contractor has taken his part ; very extended parts have fallen to the share of some, and admirably have they executed them ; but the pathway is not yet complete, and the completion of a highly curious portion of it awaits the further labours of Mr Webster of Balruddery.

A considerable portion of the rocks of this middle formation in Scotland are of a bluish-gray colour ; in Balruddery they resemble the mudstones of the Silurian

---

Radiata: see notes for pp. 259–60. No such fossils are known from Balruddery Den. Almost certainly Miller means the fruiting body of *Parka* which, on p. 145, he followed Fleming 1831 in regarding as a plant, from the raspberry-like specimens (later thought by some to be eurypterid eggs). However, vagaries of fossilisation can preserve *Parka* as a round scrap of honeycomb-like structure as in a colonial coral, or a radially wrinkled blob like a sea anemone – either way, like a 'radiate' (Bob Davidson and Lyall Anderson, pers. comm. 2015); see Figure 56; **contractor**: railways, then in the early stages of expansion, were often built by several contractors, each allocated a stretch of track.

System ; they form at Carmylie the fissile bluish-gray pavement, so well known in commerce as the pavement of Arbroath ; they occur as a hard micaceous building-stone in some parts of Fifeshire ; in others they exist as beds of friable stratified clay, that dissolve into unctuous masses where washed by the sea. Thus the base of our second pyramid is represented by a deep belt of gray. In England the formation consists, throughout its entire depth, of beds of red and green marl, with alternating beds of the nodular limestones, to which it owes its name, and with here and there an interposing band of indurated sandstone.

We pass to the upper formation. Over the belt of gray there occurs in the pyramid a second deep belt of red conglomerate and variegated sandstone, with a band of lime atop, and over the band a thick belt of yellow sandstone, with which the system terminates.*

* There still exists some uncertainty regarding the order in which these upper beds occur. Mr Duff of Elgin places the limestone band above the yellow sandstone ; Messrs Sedgwick and Murchison assign it an intermediate position between the red and yellow. The respective places of the gray and red sandstones are also disputed, and by very high authorities ; Dr Fleming holding that the gray sandstones overlie the red (see *Cheek's Edinburgh Journal* for Feb. 1831), and Mr Lyell, that the red sandstones overlie the gray (see *Elements of Geology*, pp. 99, 100). The order adopted above consorts best with the results of the writer's observations, which have, however, been restricted chiefly to the north country. He assigns to the limestone band the middle place assigned to it by Messrs Sedgwick and Murchison, and to the gray sandstone the inferior position assigned to it by Mr Lyell ; aware, however, that the latter deposit has not only a coping, but also a basement of red sandstone,—the basement forming the upper member of the lower formation.

---

indurated sandstone: at this point in the 2nd edition, Miller inserts a long sequence on Forfarshire geology and scenery taken from two later *Witness* articles, [Miller] 1841b and 1841c; upper formation: i.e. upper Old Red Sandstone; Sedgwick and Murchison: Sedgwick and Murchison 1829b, pp. 150–53; Fleming: Fleming 1831, p. 81.

Thus the second pyramid consists mineralogically, like the first, of three great divisions or bands; its two upper belts belonging, like the three belts of the other, to but one formation,—the formation known in England as the Quartzose Conglomerate. It is largely developed in Scotland. We find it spread over extensive areas in Moray, Forfar, Fife, Roxburgh, and Berwickshires. In England it is comparatively barren in fossils: the only animal organic remains yet detected in it was a single scale of the *Holoptychius* found by Mr Murchison; and though it contains vegetable organisms in more abundance, so imperfectly are they preserved, that little else can be ascertained regarding them than that they were land plants, but not identical with the plants of the Coal Measures. In Scotland the formation is richly fossiliferous, and the remains belong chiefly to the animal kingdom. It is richly fossiliferous, too, in Russia, where it was discovered by Mr Murchison, during the summer of last year, spread over areas many thousand square miles in extent. And there, as in Scotland, the *Holoptychius* is its characteristic fossil.

The fact seems especially worthy of remark. The organisms of some of the newer formations differ entirely, in widely-separated localities, from their cotemporary organisms, just as in the existing state of things the plants and animals of Great Britain differ from the plants and animals of Lapland or Sierra Leone. A geologist who has acquainted himself with the belemnites, baculites, turrilites, and sea-urchins of the Cretaceous group in England and the north of France, would discover that he had got into an en-

---

found by Mr Murchison: in Wales (Murchison 1839, part II, pp. 589, 601).

tirely new field among the hippurites, sphærulites,
and nummulites of the same formations, in Greece,
Italy, and Spain ; nor, in passing to the tertiary deposits, would he find less striking dissimilarities between the gigantic mail-clad megatherium and huge
mastodon of the Ohio and the La Plate, and the
monsters, their cotemporaries, the hairy mammoth
of Siberia, and the hippopotamus and rhinoceros of
England and the Continent. In the more ancient
geological periods, ere the seasons began, the case is
essentially different; the cotemporary formations, when
widely separated, are often very unlike in mineralogical character, but in their fossil contents they are almost always identical. In these earlier ages, the atmospheric temperature seems to have depended more
on the internal heat of the earth, only partially cooled
down from its original state, than on the earth's configuration or the influence of the sun. Hence a widely-
spread equality of climate,—a green-house equalization
of heat, if I may so speak ; and hence, too, it would
seem, a widely-spread Fauna and Flora. The green-
houses of Scotland and Sweden produce the same
plants with the green-houses of Spain and Italy; and
when the world was one vast green-house, heated
from below, the same families of plants, and the same
tribes of animals, seem to have ranged over spaces immensely more extended than those geographical circles
in which, in the present time, the same plants are
found indigenous, and the same animals native. The
fossil remains of the true Coal Measures are the same
to the westward of the Alleghany Mountains as in
New Holland, India, Southern Africa, the neighbour-

---

earth's configuration: the changing pattern of the continents and oceans through
geological time, one of the controls of climate mooted by such as Lyell in his *Principles
of Geology* (1830–33, vol. I, pp. 92–143). Miller's **equality** is correct for the absence of a
latitudinal gradient to climate, but he is also using the concept of equability (lack of
marked variation) over the annual cycle (**influence of the sun**) and over geological time
(gradual cooling) seen in Lyell's work, for instance; **New Holland**: Australia.

hood of Newcastle, and the vicinity of Edinburgh. And I entertain little doubt that, on a similar principle, the still more ancient organisms of the Old Red Sandstone will be found to bear the same character all over the world.

CHAPTER IX.

Fossils of the Upper Old Red Sandstone much more imperfectly preserved than those of the Lower.—The Causes obvious.—Difference between the two Groupes, which first
strikes the Observer, a Difference in Size.—The *Holoptychius*
a characteristic Ichthyolite of the Formation.—Description
of its huge Scales.—Of its Occipital Bones, Fins, Teeth, and
general Appearance.—Cotemporaries of the *Holoptychius*.—
Sponge-like Bodies.—Plates resembling those of the Sturgeon.—Teeth of various Forms, but all evidently the Teeth of
Fishes.—Limestone Band and its probable Origin.—Fossils
of the Yellow Sandstone.—The *Pterichthys* of Dura Den.—
Member of a Family peculiarly characteristic of the System.—No intervening Formation between the Old Red Sandstone and the Coal Measures.—The *Holoptychius* cotemporary for a time with the *Megalichthys*.—The Columns of Tubal Cain.

THE different degrees of entireness in which the
geologist finds his organic remains, depend much less
on their age than on the nature of the rock in which
they occur; and as the arenaceous matrices of the
Upper and Middle Old Red Sandstones have been
less favourable to the preservation of their peculiar
fossils than the calcareous and aluminous matrices
of the Lower, we frequently find the older organisms of the system fresh and unbroken, and the
more modern existing as mere fragments. A fish

thrown into a heap of salt would be found entire after the lapse of many years ; a fish thrown into a heap of sand would disappear in a mass of putrefaction in a few weeks; and only the less destructible parts, such as the teeth, the harder bones, and perhaps a few of the scales, would survive. Now, limestone, if I may so speak, is the preserving salt of the geological world ; and the conservative qualities of the shales and stratified clays of the Lower Old Red Sandstone are not much inferior to those of lime itself; while, in the Upper Old Red, we have merely beds of consolidated sand, and these, in most instances, rendered less conservative of organic remains than even the common sand of our shores, by a mixture of the red oxide of iron. The older fossils, therefore, like the mummies of Egypt, can be described well nigh as minutely as the existences of the present creation; the newer, like the comparatively modern remains of our churchyards, exist, except in a few rare cases, as mere fragments, and demand powers such as those of a Cuvier or an Agassiz to restore them to their original combinations.

But cases, though few and rare, do occur in which, through some favourable accident connected with the death or sepulture of some individual existence of the period, its remains have been preserved almost entire; and one such specimen serves to throw light on whole heaps of the broken remains of its cotemporaries. The single elephant, preserved in an iceberg beside the Arctic Ocean, illustrated the peculiarities of the numerous extinct family to which it belonged, and whose bones and huge tusks whiten the wastes of Si-

---

oxide: compare Lyell 1838, p. 419: 'It is a general fact, and one not yet accounted for, that scarcely any fossil remains are preserved in stratified rocks in which this oxide of iron abounds'; see also pp. 21, 191; church-yards: i.e. of the old kind, not the modern sanitised cemetery. The re-use of graves meant plenty of scattered and decaying bone fragments as well as more complete specimens – especially, one imagines, in the old kirkyards at Cromarty and Rosemarkie, on raised beaches with (probably) sandy soils where bone preservation was apt to be poor. Miller's comments doubtless reflected his experience of carving monuments there; **elephant ... iceberg**: the widely reported

beria. The human body found in an Irish bog, with the ancient sandals of the country still attached to its feet by thongs, and clothed in a garment of coarse hair, gave evidence that bore generally on the degree of civilization attained by the inhabitants of an entire district in a remote age. In all such instances the character and appearance of the individual bear on those of the tribe. In attempting to describe the organisms of the Lower Old Red Sandstone, where the fossils lie as thickly in some localities as herrings on our coasts in the fishing season, I felt as if I had whole tribes before me. In describing the fossils of the Upper Old Red Sandstone I shall have to draw mostly from single specimens. But the evidence may be equally sound so far as it goes.

The difference between the superior and inferior groupes of the system which first strikes an observer, is a difference in the size of the fossils of which these groupes are composed. The characteristic organisms of the Upper Old Red Sandstone are of much greater bulk than those of the Lower, which seem to have been characterized by a mediocrity of size throughout the entire extent of the formation. The largest ichthyolites of the group do not seem to have much exceeded two feet or two feet and a half in length; its smaller average from an inch to three inches. A jaw, in the possession of Dr Traill,—that of an Orkney species of *Holoptychius*, and by much the largest in his collection,—does not exceed in bulk the jaw of a full-grown coal-fish or cod; his largest *Coccosteus* must have been a considerably smaller fish than an ordinary-sized turbot; the largest ichthyolite found

discovery, in 1799, of a frozen mammoth carcase with part of its fleece intact in the permafrost near the Lena River in Siberia: Lyell 1830–33, vol. I, p. 99; Rudwick 2005, pp. 559–60.

by the writer was a *Diplopterus*, of, however, smaller dimensions than the ichthyolite to which the jaw in the possession of Dr Traill must have belonged; the remains of another *Diplopterus* from Gamrie, the most massy yet discovered in that locality, seem to have composed the upper parts of an individual about two feet and a half in length. The fish, in short, of the lower ocean of the Old Red Sandstone,—and I can speak of it throughout an area which comprises Orkney and Inverness, Cromarty and Gamrie, and which must have included about ten thousand square miles, —ranged in size between the stickleback and the cod; whereas some of the fish of its upper ocean were covered by scales as large as oyster-shells, and armed with teeth that rivalled in bulk those of the crocodile. They must have been fish on an immensely larger scale than those with which the system began. There have been scales of the *Holoptychius* found in Clashbennie which measure three inches in length by two and a half in breadth, and a full eighth part of an inch in thickness. There occur occipital plates of fishes in the same formation in Moray, a full foot in length by half a foot in breadth. The fragment of a tooth still attached to a piece of the jaw, found in the sandstone cliffs that overhang the Findhorn, measures an inch in diameter at the base. A second tooth of the same formation, of a still larger size, disinterred by Mr Patrick Duff from out the conglomerates of the *Scat Craig*, near Elgin, and now in his possession, measures two inches in length by rather more than an inch in diameter. There occasionally turn up in the sandstones of Perthshire ichthyodorulites that in bulk

Clashbennie: see p. 160.

and appearance resemble the teeth of a harrow, rounded at the edges by a few months' wear, and which must have been attached to fins not inferior in general bulk to the dorsal fin of an ordinary-sized porpoise. In short, the remains of a Patagonian burying-ground would scarcely contrast more strongly with the remains of that battle-field described by Addison, in which the pigmies were annihilated by the cranes, than the organisms of the upper formation of the Old Red Sandstone contrast with those of the lower.

Of this upper formation the most characteristic and most abundant ichthyolite, as has been already said, is the *Holoptychius*. The large scales and plates, and the huge teeth, belong to this genus. It was first introduced to the notice of geologists in a paper read before the Wernerian Society in May 1830, by Professor Fleming, and published by him in the February of the following year, in *Cheek's Edinburgh Journal*. Only detached scales and the fragment of a tooth had as yet been found; and these he minutely described as such, without venturing to hazard a conjecture regarding the character or family of the animal to which they had belonged. They were submitted some years after to Agassiz, by whom they were referred, though not without considerable hesitation, to the genus *Gyrolepis;* and the doubts of both naturalists serve to show how very uncertain a guide mere analogy proves to even men of the first order, when brought to bear on organisms of so strange a type as the ichthyolites of the Old Red Sandstone. At this stage, however, an almost entire specimen of the creature was discovered in the sandstones of Clash-

---

Patagonian: the natives of Patagonia were at one time supposed to be exceptionally tall; **Addison**: Joseph Addison (1672–1719), English author; one of his works in Latin, later translated by Samuel Johnson, was *Proelium inter pygmaos et grues* ('Battle between the Pygmies and the Cranes', 1698); **Wernerian Society**: founded in 1808, one of the key Edinburgh societies for natural science; **in *Cheek's Edinburgh Journal***: Fleming 1831; see also Andrews 1982a, p. 12.

bennie, by the Rev. James Noble of St Madoes, a gentleman who, by devoting his leisure hours to Geology, has extended the knowledge of this upper formation, and whose name has been attached by Agassiz to its characteristic fossil, now designated the *Holoptychius Nobilissimus.* His specimen at once decided that the creature had been no *Gyrolepis,* but the representative of a new genus not less strangely organized, and quite as unlike the existences of the present time as any existence of all the past. So marked are the peculiarities of the *Holoptychius,* that they strike the commonest observer.

The scales are very characteristic. They are massy elliptical plates, scarcely less bulky in proportion to their extent of surface than our smaller copper coin, composed internally of bone and externally of enamel, and presenting on the one side a porous structure, and on the other, when well preserved, a bright glossy surface. The upper or glossy side is the more characteristic of the two. I have placed one of them before me. Imagine an elliptical ivory counter, an inch and a half in length by an inch in breadth, and nearly an eighth part of an inch in thickness, the larger diameter forming a line which, if extended, would pass longitudinally from head to tail through the animal which the scale covered. On the upper or anterior margin of this elliptical counter, imagine a smooth selvage or border three-eighth parts of an inch in breadth. Beneath this border there is an inner border of detached tubercules, and beneath the tubercules, large undulating furrows, which stretch longitudinally towards the lower end of the ellipsis.

Noble: Rev. James Noble (1800–48), minister of St Madoes on the north bank of the Firth of Tay (Andrews 1982a, p. 62 n. 74). He collected, from Clashbennie close by his manse, the first complete specimens of *Holoptychius nobilissimus* including the fine skeleton found in July 1836 and described by Agassiz in Murchison 1839, part II, pp. 599–601, plate 2[bis], fig. 1; **scales:** see Plate VII, fig. 2.

Some of these waved furrows run unbroken and separate to the bottom, some merge into their neighbouring furrows at acute angles, some branch out and again unite, like streams which inclose islands, and some break into chains of detached tubercules. No two scales exactly resemble one another in the minuter peculiarities of their sculpture, if I may so speak, just as no two pieces of lake or sea may be roughened after exactly the same pattern during a gale ; and yet in general appearance they are all wonderfully alike. Their *style* of sculpture is the same,—a style which has sometimes reminded me of the Runic knots of our ancient north-country obelisks. Such was the scale of the creature. The head, which was small compared with the size of the body, was covered with bony plates, roughened after a pattern somewhat different from that of the scales, being tuberculated rather than ridged ; but the tubercules present a confluent appearance, just as chains of hills may be described as confluent, the base of one hill running into the base of another. The operculum seems to have been covered by one entire plate,—a peculiarity observable, as has been remarked, among some of the ichthyolites of the Lower Old Red Sandstone, such as the *Diplopterus, Dipterus,* and *Osteolepis.* And it, too, has its field of tubercules, and its smooth marginal selvage or border, on which the lower edges of the upper occipital plates seem to have rested, just as in the roof of a slated building part of the lower tier of slates are overtopped and covered by the tier above. The scales towards the tail suddenly diminish at the ventral fins to about

L

Runic knots ... obelisks: the decorative carvings found on Pictish steles, upright stones or cross-slabs (e.g. the Clach a' Charridh or Shandwick Stone to the north of the Cromarty Firth, and Sueno's Stone at Forres mentioned on p. 195). These and the Picts themselves were then often viewed as Scandinavian in origin (Huxley 1870, p. 199; Wawn 2000, p. 76), so the stones were often referred to as 'runic obelisks' by analogy or identification with Norse runestones.

one-fourth the size of those on the upper part of the
body; the fins themselves are covered at their bases,
which seem to have been thick and fleshy like the
base of the pectoral fin in the cod or haddock, with
scales still more minute; and from the scaly base the
rays diverge like the radii of a circle, and terminate
in a semicircular outline.   The ventrals are placed
nearer the tail, says Agassiz, than in any other ganoid
fish. (See Plate IX. fig. 2.)

But no such description can communicate an adequate conception to the reader of the strikingly picturesque appearance of the *Holoptychius*, as shown in
Mr Noble's splendid specimen.   There is a general
massiveness about the separate portions of the creature, that imparts ideas of the gigantic, independently
of its bulk as a whole; just as a building of moderate size, when composed of very ponderous stones,
has a more imposing effect than much larger buildings in which the stones are smaller.   The body
measures a foot across by two feet and a half in
length, exclusive of the tail, which is wanting; but
the armour in which it is cased might have served a
crocodile or aligator of five times the size.   It lies
on its back, on a mass of red sandstone; and the
scales and plates still retain their bony colour, slightly
tinged with red, like the skeleton of some animal that
had lain for years in a bed of ferruginous marl or clay.
The outline of the occipital portion of the specimen
forms a low Gothic arch, of an intermediate style between the round Saxon and the pointed Norman.
This arch is filled by two angular pane like plates, separated by a vertical line, that represents, if I may use

<hr>

Plate IX: note spelling errors, 'Lydlii' for 'lyellii' and 'Nobilissimas' for '–us'.

PLATE IX.

Cephalaspis Lydlii. Agass.

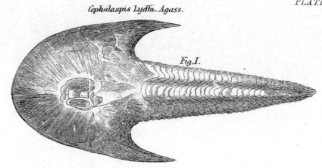

Fig. 1.

Holoptychius Nobilissimus. Agass.

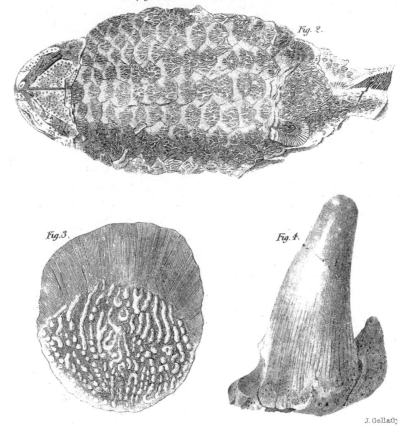

Fig. 2.

Fig. 3.

Fig. 4.

J. Gellatly

Published by John Johnstone, Hunter Square, Edinburgh.

the figure, the dividing astragal ; and the under jaw, with its two sweeping arcs or branches, constitutes the frame. All of the head which appears is that under portion of it which extends from the upper part of the belly to the snout. The belly itself is thickly covered by huge carved scales, that, from their massiveness and regular arrangement, remind one of the flags of an ancient stone roof. The carving varies as they descend towards the tail, being more in the ridged style below, and more in the tuberculated style above. So fairly does the creature lie on its back, that the ventral fins have fallen equally, one on each side, and, from their semicircular form, remind one of the two pouch-holes in a lady's apron, with their laced flaps. The entire outline of the fossil is that of an elongated ellipsis, or rather spindle, a little drawn out towards the caudal extremity. The places of all the fins are not indicated, but, as shown by other specimens, they seem to have been crowded together towards the lower extremity, like those of the *Glyptolepis*, an ichthyolite which, in more than one respect, the *Holoptychius* must have resembled, and which, from this peculiarity, presents a brush-like appearance,—the head and shoulders representing the handle, and the large and thickly-clustered fins the spreading bristles.

Some of the occipital bones of the *Holoptychius* are very curious and very puzzling. There are pieces rounded at one of the ends, somewhat in the manner of the neck-joints of our better-known quadrupeds, and which have been mistaken for vertebræ, but which present evidently, at the apparent joint, the enamel peculiar to the outer surface of all the plates and

---

astragal: architectural term (in the Scots sense), a glazing bar in a window.

scales of the creature, and which belonged, it is probable, to the snout. There are saddle-shaped bones
too, which have been regarded as the central occipital
plates of a new species of *Coccosteus*, but whose style
of confluent tubercule belongs evidently to the *Holoptychius*. The jaws are exceedingly curious. They
are composed of as solid bone as we usually find in
the jaws of mammalia; and the outer surface, which
is covered, in animals of commoner structure, with
portions of the facial integuments, we find polished
and japanned, and fretted into tubercules. The jaws
of the creature, like those of the *Osteolepis* of the lower formation, were naked jaws: it is indeed more than
probable that all its real bones were so, and that the
internal skeleton was cartilaginous. A row of thickly-set pointed teeth ran along the japanned edges of
the mouth,—what in fish of the ordinary construction would be the lips; and inside this row there was
a second and widely-set row of at least twenty times
the bulk of the other, and which stood up over and
beyond it, like spires in a city over the rows of lower
buildings in front. A nearly similar disposition of teeth
seems also to have characterized the *Megalichthys*, but
the contrast in size was somewhat less marked. One
of the most singularly-formed bones of the formation
will be found, I doubt not, when perfect specimens of
the upper part of the creature shall be procured, to
have belonged to the *Holoptychius*. It is a huge ichthyodorulite, formed, box-like, of four nearly rectan-
gular planes, terminating in a point, and ornamented on two of the sides by what in a work of art the
reader would at once term a species of Chinese fret-

work.　Along the centre there runs a line of lozenges, slightly truncated where they unite, just as, in plants that exhibit the cellular texture, the lozenge-shaped cells may be said to be truncated.　At the sides of the central line there run lines of half lozenges, which occupy the space to the edges.　Each lozenge is marked by lines parallel to the lines which describe it, somewhat in the manner of the plates of the tortoise.　The centre of each is thickly tuberculated ; and what seems to have been the anterior plane of the ichthyodorulite is thickly tuberculated also, both in the style of the occipital plates and jaws of the *Holoptychius*.　This curious bone, which seems to have been either hollow inside, or, what is more probable, filled with cartilage, measures in some of the larger specimens an inch and half across at the base on its two broader planes, and rather more than half an inch on its two narrower ones.

Geologists have still a great deal to learn regarding the cotemporaries of the *Holoptychius Nobilissimus*. The lower portion of that upper formation to which it more especially belongs,—the portion represented in our second pyramid by the conglomerate and sandstone bar,—though unfavourable to the preservation of animal remains, represents assuredly no barren period. It has been found to contain organic bodies, that vary in shape like the sponges of our existing seas, which in general appearance they somewhat resemble, but whose class, and even kingdom, is yet to fix.　It contains, besides, in considerable abundance, though in a state of very imperfect preservation, scales that differ from those of the *Holoptychius* and from one another.

---

curious bone: of a fish later named *Placothorax* by Agassiz: Miller 1858a, p. 182 n.; **bodies ... like sponges**: discussed in detail in a later edition, in which Miller wondered if they were remains of living organisms (**organic**) at all, and spotted corrugations running along each one's longitudinal axis, signs of 'some external cause acting on the whole in one direction' (Miller 1858a, p. 182 n.). An acute observation, for they were almost certainly rip-up clasts, common in the Old Red Sandstone in this area (MacGregor 1996, p. 35). These are flakes of consolidated mud or shale ripped up by a sudden water current, and transported elsewhere, perhaps into a contrasting sandy sediment.

One of these, figured and described by Professor Fleming in *Cheek's Edinburgh Journal*, bearing on its upper
surface a mark like a St Andrew's cross, surrounded
by tuberculated dottings, and closely resembling in
external appearance some of the scales of the common
sturgeon, " may be referred, with some probability,"
says the Professor, " to an extinct species of the genus *Accipenser*.* The deposit too abounds in teeth,

* May I crave the attention of the reader to a brief statement of fact ? I have said that Professor Fleming, when he
minutely described the scales of the *Holoptychius*, hazarded
no conjecture regarding the generic character of the creature
to which they had belonged ; he merely introduced them to
the notice of the public as the scales of some " vertebrated animal, probably those of a fish." I now state that he described
the scales of a cotemporary ichthyolite as bearing in external
appearance a " close resemblance to some of the scales of the
common sturgeon." It has been asserted, that it was the
scales of the *Holoptychius* which he thus described, " referring them to an extinct species of the genus *Accipenser ;*"
and the assertion has been extensively credited, and by some
of our highest geological authorities. Agassiz himself, evidently in the belief that the Professor had fallen into a palpable error, deems it necessary to prove that the *Holoptychius*
could have borne " no relation to the *Accipenser* or sturgeon."
Mr Murchison, in his *Silurian System*, refers also to the supposed mistake. The person with whom the misunderstanding
seems to have originated is the Rev. Dr Anderson of Newburgh. About a twelvemonth after the discovery of Professor
Fleming in the sandstones of Drumdryan, a similar discovery
was made in the sandstones of Clashbennie by a geologist of
Perth, who, on submitting his new-found scales to Dr Anderson, concluded, with the Doctor, that they could not be
other than *oyster shells ;* though, eventually, on becoming
acquainted with the decision of Professor Fleming regarding
them, both gentlemen were content to alter their opinion, and
to regard them as scales. The Professor, in his paper on
the Old Red Sandstone in *Cheek's Journal*, referred incident-

Fleming: the quotation is paraphrased from Fleming 1831, p. 86; **brief statement**: for
discussion of the affair narrated in this enormous footnote, see Additional Note for
p. 173; **Dr Anderson**: Rev. Dr John Anderson (1796–1864), minister of St Katherine's at
Newburgh in Fife, keen geologist and antiquary: Andrews 1982a, p. 58 n. 32; **Agassiz** and
Murchison: in Murchison 1839, part II, p. 600; **paper … in** *Cheek's Journal*: Fleming
1831.

various enough in their forms to indicate a corresponding variety of families and genera among the ichthyolites to which they belonged. Some are nearly

ally to the *oyster shells* of Clashbennie,—a somewhat delicate subject of allusion; and in Dr Anderson's paper on the same formation, which appeared about seven years after in the *New Journal* of Professor Jameson, the geological world was told for the first time, that Professor Fleming had described a scale of Clashbennie *similar to those of Drumdryan*, *i. e.* those of the *Holoptychius*, as bearing a " close resemblance to some of the scales on the common sturgeon," and as probably referrible to some " extinct species of the genus *Acipenser*." Now, Professor Fleming, instead of stating that the scales were at all. similar, had stated very pointedly that they were entirely different; and not only had he *described* them as different, but he had also *figured* them as different, and had placed the figures side by side, that the difference might be the better seen. To the paper of the Professor, which contained this statement, and to which these figures were attached, Dr Anderson referred, as "read before the Wernerian Society;"—he quoted from it in the Professor's words,—he drew some of the more important facts of his own paper from it,—in his late Essay on the Geology of Fife he has availed himself of it still more largely, though with no acknowledgment, —it has constituted, in short, by far the most valuable of all his discoveries in connection with the Old Red Sandstone, and apparently the most minutely examined; and yet so completely did he fail to detect Professor Fleming's carefully-drawn distinction between the scales of the *Holoptychius* and those of its cotemporary, that when Agassiz, misled apparently by the Doctor's own statement, had set himself to show that the scaly giant of the formation could have been no sturgeon, the Doctor had the passage in which the naturalist established the fact transferred into a Fife newspaper, with, of course, the laudable intention of preventing the Fife public from falling into the *absurd mistake* of Professor Fleming. There seems to be something rather inexplicable in all this; but there can be little doubt Dr Anderson could satisfactorily explain the whole matter without once referring to the *oyster*

Dr Anderson's paper: Anderson 1837, p. 138; late Essay: Anderson 1841, made separately and publicly available in or by September 1840: see p. 174; [Miller] 1840h; Andrews 1982a, p. 62 n. 73; Fife newspaper: untraced; see Additional Note for p. 173.

straight, like those of the *Megalichthys* of the Coal Measures; some are bent, like the beak of a hawk or eagle, into a hook-form; some incline first in one di-

shells of Clashbennie. It is improbable that he could have wished or intended to injure the reputation of a gentleman to whose freely-imparted instructions he is indebted for by much the greater portion of his geological skill,—whose remarks, written and spoken, he has so extensively appropriated in his several papers and essays,—and whose character is known far beyond the limits of his country, for untiring research, philosophic discrimination, and all the qualities which constitute a naturalist of the highest order. Dr Johnston of Berwick, in his *History of British Zoophites* (a work of an eminently scientific character), justly " ascribes to the labours and writings" of Professor Fleming " no small share in diffusing that taste for Natural History which is now abroad." And as an interesting corroboration of the fact I may state, that Dr Malcolmson of Madras lately found an elegant Italian translation of *Fleming's Philosophy of Zoology* high in repute among the elite of Rome. Lest it should be supposed I do Dr Anderson injustice in these remarks, I subjoin the grounds of them in the following extracts from Professor Fleming's paper in *Cheek's Journal*, and from the paper in *Jameson's New Edinburgh Journal*, in which the Doctor purports to give a digest of the former, without once referring, however, to the periodical in which it is to be found.

" In the summer of 1827," says Dr Fleming, " I obtained from Drumdryan quarry, to the south of Cupar, situate in the higher strata of yellow sandstone, certain organisms, which I readily referred to the scales of vertebrated animals, probably those of a fish. The largest (see Plate II. fig 1, '*figure of a scale of the Holoptychius'*) was one inch and one-tenth in length, about one inch and two-tenths in breadth, and not exceeding the fiftieth of an inch in thickness. The part which, when in its natural position, had been embedded in the cuticle, is comparatively smooth, exhibiting, however, in a very distinct manner the semicircularly-parallel layers of growth, with obsolete diverging striæ, giving to the surface, when under a lens, a reticulated aspect. The part naturally exposed is

Dr Johnston: George Johnston (1797–1855) of Berwick-upon-Tweed, important botanist and especially marine biologist (Davis 1995). The reference is to Johnston 1838, p. 206, which actually says 'considerable share'; extracts: Fleming 1831, pp. 84–86. Miller has added capitalisation and, in italics, an explanatory interpolation of a drawing which he does not reproduce.

rection and then in the opposite one, like nails that have been drawn out of a board by the carpenter at two several wrenches, and bent in opposite angles at

marked with longitudinal, waved, rounded, anastomosing ridges, which are smooth and glossy. The whole of the inside of the scale is smooth, though exhibiting with tolerable distinctness the layers of growth. The form and structure of the object indicated plainly enough that it had been a scale, a conclusion confirmed by the detection of the phosphate of lime in its composition. At this period I inserted a short notice of the occurrence of these scales in our provincial newspaper, the *Fife Herald*, for the purpose of attracting the attention of the workmen and others in the neighbourhood, in order to secure the preservation of any other specimens which might occur.

" Nearly a year after these scales had been discovered, not only in the upper, but even in some of the lower beds of the Yellow Sandstone, I was informed that *oyster shells* had been found in a quarry, in the Old Red Sandstone, at Clashbennie, near Errol, in Perthshire, and that specimens were in the possession of a gentleman in Perth. Interested in the intelligence, I lost no time in visiting Perth, and was gratified to find that the supposed oyster shells were in fact similar to those which I had ascertained to occur in a higher part of the series. The scales were, however, of a larger size, some of them exceeding three inches in length and one-eighth of an inch in thickness. Upon my visit to the quarry I found the scales, as in the Yellow Sandstone, most abundant in those parts of the rock which exhibited a brecciated aspect. Many patches a foot in length, full of scales, have occurred ; but as yet no entire impression of a fish has been obtained.   *   *   *

" Another scale, DIFFERING FROM THOSE ALREADY NOTICED (see Plate II. fig. 3, '*figure of an oblong tuberculated plate traversed diagonally by lines, which, bisecting one another a little above the centre, resembles a St Andrew's cross, and marked on the edges by faintly radiating lines'),* is about an inch and a quarter in length and an inch in breadth. In external appearance it bears a very close resemblance to some of the scales on the common sturgeon, and may, with some probability, be

---

short notice: not traced.

each wrench; some are bulky and squat, some long and slender; and in almost all the varieties, whether curved or straight, squat or slim, the base is elegantly striated like the flutings of a column.   In the splendid specimen found in the sandstones of the Findhorn, the tooth is still attached to a portion of the jaw, and shows, from the nature of the attachment, that the creature to which it belonged must have been a true fish, not a reptile.   The same peculiarity is observable in two other very fine specimens in the collection of Mr Patrick Duff of Elgin.   Both in saurians and in toothed cetaceæ, such as the porpoise, the teeth are inserted in sockets.   In the ichthyolites of this formation, so far as these are illustrated by its better specimens, the teeth, as in existing fish, are merely placed flat upon the jaw, or in shallow pits, which seem almost to indicate that the contrivance of sockets might be afterwards resorted to.   Immediately over

referred to an extinct species of the genus *Accipenser*."— *Cheek's Edinburgh Journal*, Feb. 1831, p. 85.

" Dr Fleming in 1830," says Dr Anderson, " read before the Wernerian Society a notice ' on the occurrence of scales of vertebrated animals in the Old Red Sandstone of Fifeshire.' These organisms, as described by him, occurred in the Yellow Sandstone of Drumdryan and the Gray Sandstone of Parkhill. From the former locality scales of a fish were obtained.   *   * The same paper (Professor Fleming's) contains a notice of SIMILAR SCALES in the Old Red Sandstone of Clashbennie, near Errol, in Perthshire, ONE OF WHICH is described as bearing ' a very close resemblance to some of the scales on the common sturgeon, and may with some probability be referred to an extinct species of the genus *Accipenser*.' "— *Prof. Jameson's Edin. New Phil. Journal*, Oct. 1837, p. 138.

---

splendid specimen: see p. 158; **Anderson**: Anderson 1837, p. 138.

the sandstone and conglomerate belt in which these organisms occur there rests, as has been said, a band of limestone, and over the limestone a thick bed of yellow sandstone, in which the system terminates, and which is overlaid in turn by the lower beds of the carboniferous group.

The limestone band is unfossiliferous, and resembling in mineralogical character the Cornstones of England and Wales. It has been described as the Cornstone of Scotland; but the fact merely furnishes one illustration of many, of the inadequacy of a mineralogical nomenclature for the purposes of the geologist. In the neighbourhood of Cromarty the lower formation abounds in beds of nodular limestone, identical in appearance with the Cornstone;—in England similar beds occur so abundantly in the middle formation that it derives its name from them;—in Fife they occur in the upper formation exclusively. Thus the formation of the *Coccosteus* and *Dipterus* is a Cornstone formation in the first locality; that of the *Cephalaspis* and the gigantic lobster in the second; that of the *Holoptychius Nobilissimus* in the third. We have but to vary our field of observation to find all the formations of the system *Cornstone formations* in turn. The limestone band of the upper member presents exactly similar appearances in Moray as in Fife. It is in both of a yellowish-green or gray colour, and a concretionary structure, consisting of softer and harder portions, that yield so unequally to the weather, as to exhibit in exposed cliffs and boulders a brecciated aspect, as if it had been a mechanical, not a chemical deposit; though

---

mineralogical: mineralogical or lithological similarity between two rocks is not a reliable guide to whether they are of the same age (see also pp. 184–85); **Wales. It has**: should read 'Wales it has' (author's erratum, emended in 2nd edition); **chemical**: i.e. the rock is, at least significantly, formed by precipitation of dissolved matter rather than by the accumulation and deposition of sediment.

its origin must unquestionably have been chemical.
It contains minute crystals of galena, and abounds
in masses of a cherty, siliceous substance, that strike
fire with steel, and which, from the manner in which
they are incorporated with the rock, show that they
must have been formed along with it. From this
circumstance, and from the general resemblance it
bears to the deposits of the thermal waters of volcanic districts which precipitate siliceous mixed with
calcareous matter, it has been suggested, and by no
mean authority, that it must have derived its origin
from hot springs. The bed is several yards in thickness; and as it appears both in Moray and in Fife,
in localities at least a hundred and twenty miles
apart, it must have been formed, if formed at all in
this manner, at a period when the volcanic agencies
were in a state of activity at no great distance from
the surface.

The upper belt of yellow stone, the terminal layer
of the pyramid, is fossiliferous both in Moray and
Fife,—more richly so in the latter county than even
the conglomerate belt that underlies it, and its organisms are better preserved. It was in this upper
layer, in Drumdryan quarry, to the south of Cupar,
that Professor Fleming found the first-discovered
scales of the *Holoptychius.* At Dura Den, in the same
neighbourhood, a singularly rich deposit of animal remains was laid open a few years ago by some workmen, when employed in excavating a water-course for
a mill. The organisms lay crowded together, a single
slab containing no fewer than thirty specimens, and
all in a singularly perfect state of preservation. The

whole space excavated did not exceed forty square yards in extent, and yet in these forty yards there were found several genera of fishes new to Geology, and not yet figured nor described,—a conclusive proof in itself that we have still very much to learn regarding the fossils of the Old Red Sandstone. By much the greater portion of the remains disinterred on this occasion were preserved by a lady in the neighbourhood, and the news of the discovery spreading over the district, the Rev. Dr Anderson of Newburgh was fortunately led to discover them anew in her possession. The most abundant organism of the group was a variety of *Pterichthys*,—the sixth species of this very curious genus now discovered in the Old Red Sandstones of Scotland; and as the Doctor had been lucky enough to find out for himself some years before, that the scales of the *Holoptychius* were oyster-shells, he now ascertained, with quite as little assistance from without, that the *Pterichthys* must have been surely a huge beetle. As a beetle, therefore, he figured and described it in the pages of a Glasgow topographical publication, *Fife Illustrated*. True, the characteristic elytra were wanting, and some six or seven tuberculated plates substituted in their room; nor could the artist, with all his skill, supply the creature with more than two legs; but ingenuity did much for it notwithstanding; and by lengthening the snout, insect-like, into a point,—by projecting an eye, insect-like, on what had mysteriously grown into a head,—by rounding the body, insect-like, until it exactly resembled that of the large " twilight shard,"—by exaggerating the tubercules seen in profile on the paddles until

---

Fife Illustrated: Anderson 1840; **Anderson**: see Additional Note for p. 173 and Figure 54; **elytra**: the rigid wing-covers of a beetle, held out in flight; **'twilight shard'**: in Shakespeare's *Macbeth* (Act III, Scene 2) Macbeth alludes to the twilight flight of the dung-beetle or perhaps cockchafer: 'ere to black Hecate's summons / The shard-borne beetle with his drowsy hums / Hath rung night's yawning peal'. The beetle's wing-cover is convex and shiny like a pottery shard. Perhaps Miller was also thinking of another twilight reference, 'the beetle wheels his droning flight', in Gray's 'Elegy Written in a Country Churchyard'. For other allusions to *Macbeth*, see pp. 41, 215.

they stretched out, insect-like, into bristles,—and by
carefully sinking the tail, which was not insect-like,
and for which no possible use could be discovered at
the time,—the Doctor succeeded in making the *Pterichthys* of Dura Den a very respectable beetle indeed.
In a later publication, an Essay on the Geology of
Fifeshire, which appeared in September last in the
Quarterly Journal of Agriculture, he states, after referring to his former description, that among the
higher geological authorities some were disposed to
regard the creature as an extinct crustaceous animal,
and some as belonging to a tribe closely allied to the
Chelonia. Agassiz, as the writer of these chapters ven-
tured some months ago to predict, has since pronounced
it a fish,—a *Pterichthys* specifically different from the
five varieties of this ichthyolite which occur in the
lower formation of the system, but generically the
same. I very lately enjoyed the pleasure of examining the *bona fide* ichthyolite itself,—one of the specimens of Dura Den, and apparently one of the more
entire, in the collection of Professor Fleming. Its
character as a *Pterichthys* I found very obvious ; but
neither the Professor nor myself were ingenious
enough to discover in it any trace of the beetle of Dr
Anderson.

Is it not interesting to find this very curious genus in both the lowest and highest fossiliferous beds
of the system, and constituting, like the *Trilobite* genus of the Silurian group, its most characteristic organisms ? The *Trilobite* has a wide geological range,
extending from the upper Cambrian rocks to the
upper Coal Measures. But though the range of the

September last: see note for p. 167; ***Quarterly Journal of Agriculture***: evidently
Anderson 1841, as corroborated by the mention of the publisher in the original *Witness*
article, [Miller] 1840h.

genus is wide, that of every individual species of which it consists is very limited. The *Trilobites* of the upper Coal Measures differ from those of the Mountain Limestone, these again from the *Trilobites* of the upper Silurian strata, these yet again from the *Trilobites* of the underlying middle beds, and these from the *Trilobites* that occur in the base of the system. Like the coins and medals of the antiquary, each represents its own limited period, and the whole taken together yield a consecutive record. But while we find them merely scattered over the later formations in which they occur, and that very sparingly, in the Silurian System we find them congregated in such vast crowds, that their remains enter largely into the composition of many of the rocks which compose it. The *Trilobite* is the distinguishing organism of the group, marrying, if I may so express myself, its upper and lower beds ; and what the *Trilobite* is to the Silurian formations the *Pterichthys* seems to be to the formations of the Old Red Sandstone,—with this difference, that, so far as is yet known, it is restricted to this system alone, occurring in neither the Silurian System below, nor in the Coal Measures above.

I am but imperfectly acquainted with the localities in which the upper beds of the Old Red Sandstone underlie the lower beds of the Coal Measures, or where any gradation of character appears. The upper yellow sandstone belt is extensively developed in Moray, but it contains no trace of carbonaceous matter in even its higher strata, and no other remains than those of the *Holoptychius* and its cotemporaries. The system in the north of Scotland differs as much from the carboniferous group in its upper as in its lower rocks; and a similar difference has been remarked in Fife, where the groupes appear in contact a few miles to the west of St Andrews. In England, in repeated instances, the junction, as shown by Mr Murchison, in singularly instructive sections, is well marked, the carboniferous limestones resting conformably on the Upper Old Red Sandstone. No other system interposed between them; and it is corroborative of the fact, that in Russia, scales of the *Megalichthys,* a fish of the Coal Measures, were found by Mr Murchison in an upper bed of the Old Red Sandstone, mingled with those of the *Holoptychius;* and that in Burdiehouse,—a lower bed of the Coal Measures,—scales of the *Holoptychius* have been found mingled with those of the *Megalichthys.* Both ichthyolites cross, as it were, the borders of their respective formations; the one witnessed the closing twilight of the more ancient system, the other the dawn of the system which succeeded it. They were cotemporaries for a time, somewhat in the manner that Shem was cotemporary with Isaac.

There is a Rabbinical tradition that the sons of Tubal Cain, taught by a prophet of the coming deluge, and unwilling that their father's arts should be lost in it to posterity, erected two obelisks of brass, on which they inscribed a record of his discoveries, and that thus the learning of the family survived the cataclysm. The flood subsided, and the obelisks, sculptured from pinacle to base, were found fast fixed in the rock. Now the twin pyramids of the Old Red Sandstone, with their party-coloured bars and their thickly-crowded

---

Shem … Isaac: according to the Bible, Shem was several centuries old when Isaac was born (Genesis 11:10–27 and 21:5). Miller's geohistory thus allows for gradual as well as sudden faunal change, and is not 'catastrophist' in any strict sense of the term;
Rabbinical tradition: the 'tradition' cannot be traced in this form (see Ginzberg 2003, vol. I, pp. 115–17). It is probably based on the legend of two inscribed pillars, one of stone and the other of brick, which the children of Seth built to record their astronomical learning before the Deluge (Noah's Flood) came to destroy them. This legend is related in the *Antiquities of the Jews* by the first-century Romano-Jewish scholar Flavius

inscriptions, belong to a period immensely more remote than that of the columns of the antediluvians,
and they bear a more certain record.   I have, perhaps, dwelt too long on their various compartments;
but the Artist by whom they have been erected, and
who has preserved in them so wonderful a chronicle
of his earlier works, has willed surely that they should
be read, and I have perused but a small portion of
the whole.   Years must pass ere the entire record
can be deciphered; but of all its curiously-inscribed
sentences the result will prove the same,—they will
be all found to testify of the Infinite Mind.

Josephus (book 1, chapter 2), widely known in William Whiston's 1737 translation.
Miller, writing from memory, might have muddled this story with the decorated brass
pillars set up by Hiram before the Temple in Jerusalem (Bible, 1 Kings 7:13–22), and
substituted Tubal-Cain for Seth because of Tubal-Cain's association with metalwork in
Jewish and Christian legend.

CHAPTER X.

Speculations in the Old Red Sandstone, and their Character.—
George, first Earl of Cromarty.—His Sagacity as a Naturalist at fault in one instance.—Sets himself to dig for
Coal in the Lower Old Red Sandstone.—Discovers a fine
Artesian Well.—Value of Geological Knowledge in an economic view.—Scarce a Secondary Formation in the Kingdom in which Coal has not been sought for.—Mineral Springs
of the Old Red Sandstone.—Strathpeffer.—Its Peculiarities
whence derived.—Chalybeate Springs of Easter Ross and
the Black Isle.—Petrifying Springs.—Building-Stone and
Lime of the Old Red Sandstone.—Its various Soils.

THERE has been much money lost, and a good deal
won, in speculations connected with the Old Red
Sandstone. The speculations in which money has
been won have consorted, if I may so speak, with
the character of the system, and those in which
money has been lost have not. Instead, however, of
producing a formal chapter on the economic uses to
which its various deposits have been applied, or the
unfortunate undertakings which an acquaintance with
its Geology would have prevented, I shall throw together, as they occur to me, a few simple facts illustrative of both.

George, first Earl of Cromarty, seems, like his

George, first Earl: Sir George Mackenzie (1630–1714), latterly Viscount Tarbat and
then Earl of Cromartie, historian, Classical scholar and Restoration statesman.

McKenzie (p. 179): Sir George Mackenzie (1636–91), the Lord Advocate (chief legal
adviser of the Crown). He was **too celebrated**, i.e. infamous, as the Bluidy Mackenzie
who persecuted the Covenanters whom Miller, like many other Scots, saw as heroic
fighters against the oppressive Stuarts. Mackenzie is the anti-hero of *Old Mortality*,
Scott's novel of the 1679 Covenanter rising and its defeat at Bothwell Brig. The town

THE OLD RED SANDSTONE.

namesake and cotemporary, the too celebrated Sir George M'Kenzie of Roseavoch, to have been a man of an eminently active and inquiring mind. He found leisure, in the course of a very busy life, to write several historical dissertations of great research, and a very elaborate *Synopsis Apocalyptica*. He is the author, too, of an exceedingly curious letter on the " Second Sight," addressed to the philosophic Boyle, which contains a large amount of amusing and extraordinary fact; and his description of the formation of a peat-moss in the central Highlands of Ross-shire has been quoted by almost every naturalist who, since the days of the sagacious nobleman, has written on the formation of peat. His life was extended to extreme old age; and as his literary ardour remained undiminished till the last, some of his writings were produced at a period when most other men are sunk in the incurious indifferency and languor of old age. And among these later productions are his remarks on peat. He relates, that when a very young man, he had marked, in passing on a journey through the central Highlands of Ross-shire, a wood of very ancient trees,—doddered and moss-grown, and evidently passing into a state of death through the last stages of decay. He had been led by business into the same district many years after, when in middle life, and found that the wood had entirely disappeared, and that the heathy hollow which it had covered was now occupied by a green and stagnant morass, unvaried in its tame and level extent by either bush or tree. In his old age, he again visited the locality, and saw the green surface roughened

---

nearest his Roseavoch estate is named Avoch but pronounced 'auch', whence Miller's spelling and the modern 'Rosehaugh'; *Synopsis Apocalyptica*: a book (1707) containing explanations of biblical prophecies; '**Second Sight**': the legendary ability to see beyond the limits of space and time, often associated with Gaels; see also Miller 1835, 1850b, 1994; **letter**: S. Pepys 1828, vol. V, pp. 265–73; see Miller 1850b, pp. 151–52, 1994, p. 147; **Boyle**: Robert Boyle (1627–91), Irish natural philosopher, experimentalist and leading member of the Royal Society of London; **peat**: Rennie 1807–10, vols. I, p. 65, and II, p. 257; see Miller 1850b, p. 151, 1994, p. 147.

with dingy-coloured hollows, and several Highland-
ers engaged in it in cutting peat in a stratum several
feet in depth. What he had once seen an aged forest had now become an extensive peat-moss.

Some time towards the close of the seventeenth
century he purchased the lands of Cromarty, where
his turn for minute observation seems to have anticipated, little, however, to his own profit, some of the
later geological discoveries. There is a deep wooded
ravine in the neighbourhood of the town, traversed
by a small stream, which has laid bare, for the space
of about forty yards in the opening of the hollow, the
gray sandstone and stratified clays of the inferior fish
bed. The locality is rather poor in ichthyolites,
though I have found in it, after minute search, a
few scales of the *Osteolepis*, and on one occasion
one of the better marked plates of the *Coccosteus ;*
but in the vegetable impressions peculiar to the formation it is very abundant. These are invariably
carbonaceous, and are not unfrequently associated
with minute patches of bitumen, which in the harder
specimens present a coal-like appearance ; and the
vegetable impressions and the bitumen seem to have
misled the sagacious nobleman into the belief that
coal might be found on his new property. He accordingly brought miners from the south, and set
them to bore for coal in the gorge of the ravine.
Though there was probably a register kept of the various strata through which they passed, it must have
long since been lost ; but from my acquaintance with
this portion of the formation, as shown in the neighbouring sections, where it lies uptilted against the

---

ravine: Old Chapel Den, just below St Regulus's Chapel; **He accordingly brought
miners**: probably in fact commissioned by a later landowner, George Ross, in the mid-
eighteenth century: Alston 2006, p.188; **gorge**: see p.129.

granite gneiss of the Sutors, I think I could pretty nearly restore it. They would first have had to pass for about thirty feet through the stratified clays and shales of the ichthyolite bed, with here and there a thin band of gray sandstone, and here and there a stratum of lime; they would next have had to penetrate through from eighty to a hundred feet of coarse red and yellow sandstone,—the red greatly predominating. They would then have entered the great conglomerate, the lowest member of the formation; and in time, if they continued to urge their fruitless labours, they would arrive at the primary rock, with its belts of granite, and its veins and huge masses of hornblende. In short, there might be some possibility of their penetrating to the central fire, but none whatever of their ever reaching a vein of coal. From a curious circumstance, however, they were prevented from ascertaining, by actual experience, the utter barrenness of the formation.

Directly in the gorge of the ravine, where we may see the partially-wooded banks receding as they ascend from the base to the centre, and then bellying over from the centre to the summit, there is a fine chalybeate spring, surmounted by a dome of hewn stone. It was discovered by the miners, when in quest of the mineral which they did not and could not discover, and forms one of the finest specimens of a true Artesian well which I have anywhere seen. They had bored to a considerable depth, when, on withdrawing the kind of augre used for the purpose, a bolt of water, which occupied the whole diameter of the bore, came rushing after like the jet of a foun-

---

spring: now the Coalheugh Well, still active today. An artesian well or spring, natural or artificial, was explained for instance by Buckland (1836, vol. I, pp. 560–69). The pressure of the groundwater is greater than the head of water at the surface so that the water spouts naturally upwards, or, in the Coalheugh, forces itself up and out through the outlet on the side of the covering dome. In Miller's time, boring for water was an increasingly important technique, especially where surface water was scarce or polluted, but it needed an understanding of geology (Torrens 2003; Mather 2004).

tain, and the work was prosecuted no further; for, as
steam-engines were not yet invented, no pit could
have been wrought with so large a stream issuing into it; and as the volume was evidently restricted by
the size of the bore, it was impossible to say how
much greater a stream the source might have supplied.   The spring still continues to flow towards the
sea between its double row of cresses, at the rate of
about a hogshead per minute,—a rate considerably diminished, it is said, from its earlier volume, by some
obstruction in the bore.   The waters are not strongly
tinctured,—a consequence, perhaps, of their great
abundance ; but we may see every pebble and stalk in
their course enveloped by a ferruginous coagulum, resembling burnt sienna, that has probably been disen-
gaged from the dark red sandstone below, which is
known to owe its colour to the oxide of iron.   A
Greek poet would probably have described the inci-
dent as the birth of the Naiad ; in the north, however, which in an earlier age had also its Naiads,
though, like the fish of the Old Red Sandstone, they
have long since become extinct, the recollection of it
is merely preserved by tradition, as a curious, though
by no means poetical fact, and by the name of the
well, which is still known as the well of the *coal-
heugh*,—the old Scotch name for a coal-pit.   Calderwood tells us, in his description of a violent tempest
which burst out immediately as his persecutor James
VI. breathed his last, that in the south of Scotland
the sea rose high upon the land, and that many
" *coal heughs were drowned.*"

There is no science whose value can be adequately

---

hogshead: a variable measure, here probably the newly standardised imperial unit of
52.5 gallons (*c.* 240 litres); cresses: watercress; Naiad: in Greek mythology, a water-spirit
or nymph of a spring or river; in some versions, a naiad lived and died with its body of
water. Perhaps from Hibbert (1835, p. 274) who mocked Robert Jameson's refusal to believe that the Burdiehouse Limestone was a freshwater deposit as 'the objection, that the
deposit … was less under the dominion of the Naiads than of Neptune', Neptune being
god of the sea; Calderwood: David Calderwood (1575–1651), historian and Presbyterian
divine; from his *Historie of the Kirk of Scotland,* published (abridged) in 1646.

estimated by economists and utilitarians of the lower order. Its true quantities cannot be represented by arithmetical figures or monetary tables; for its effects on mind must be as surely taken into account as its operations on matter, and what it has accomplished for the human intellect, as certainly as what it has done for the comforts of society or the interests of commerce. Who can attach a marketable value to the discoveries of Newton? I need hardly refer to the often-quoted remark of Johnson: the beauty of the language in which it is couched has rendered patent to all, the truth which it conveys. " Whatever withdraws us from the power of the senses," says the moralist,—" whatever makes the past, the distant, or the future, predominate over the present,— advances us in the dignity of thinking beings." And Geology, in a peculiar manner, supplies to the intellect an exercise of this ennobling character. But it has also its cash value. The time and money squandered in Great Britain alone in searching for coal in districts where the well-informed Geologist could have at once pronounced the search hopeless, would much more than cover the expense at which geological research has been prosecuted throughout the world. There are few districts in Britain occupied by the secondary deposits in which at one time or another the attempt has not been made. It has been the occasion of enormous expenditure in the south of England among the newer formations, where the coal, if it at all occurs (for we occasionally meet with wide gaps in the scale), must be buried at an inapproachable depth. It led in Scotland,—in the northern

---

Newton: Sir Isaac Newton (1642–1727) set out the laws of physical dynamics and gravitation; Johnson: the quotation adapts Johnson's reflections on historical monuments on Iona in *A Journey to the Western Islands of Scotland* (1775). In the original passage, Johnson views Iona as, simultaneously, a specific place with strong local associations and a representative of universal, timeless principles (Wiltshire 1997, p. 222) – a double vision which also characterises Miller's attitude towards Cromarty in this book; searching for coal: see p. A72; secondary: in this context, sedimentary rocks.

county of Sutherland,—to an unprofitable working for many years of a sulphureous lignite of the inferior Oolite, far above the true Coal Measures. The attempt I have just been describing was made in a locality as far beneath them. There is the scene of another and more modern attempt in the same district on the shores of the Moray Frith, in a detached patch of Lias, where a fossilized wood would no doubt be found in considerable abundance, but no continuous vein even of lignite. And it is related by Dr Anderson of Newburgh, that a fruitless and expensive search after coal has lately been instituted in the Old Red Sandstone beds which traverse Strathearn and the Carse of Gowrie, in the belief that they belong, not to the Old, but to the *New* Red Sandstone,—a formation which has been successfully perforated in prosecuting a similar search in various parts of England. All these instances—and there are hundreds such—show the economic importance of the study of fossils. The Oolite has its veins of apparent coal on the coast of Yorkshire, and its still more amply developed veins—one of them nearly four feet in thickness—on the eastern coast of Sutherlandshire; the Lias has its coniferous fossils in great abundance, some of them converted into a lignite which can scarce be distinguished from a true coal; and the bituminous masses of the Lower Old Red, and its carbonaceous markings, appear identical, to an unpractised eye, with the impressions on the carboniferous sandstones, and the bituminous masses which they, too, are occasionally found to enclose. Nor does the mineralogical character of its middle beds differ in many cases from

---

Sutherland: at Brora (Owen 1995; Trewin and Hurst 2009); **more modern attempt**: see p. 185; **Anderson**: probably Anderson 1841, pp. 384–85.

that of the lower members of the New Red Sandstone. I have seen the older rock in the north of
Scotland as strongly saliferous as any of the newer
sandstones,—of well nigh as bright a brick-red tint,—
of as friable and mouldering a texture,—and variegated as thickly with its specks and streaks of green
and buff-colour. But in all these instances there are
strongly-characterized groupes of fossils, which, like
the landmarks of the navigator, or the findings of his
quadrant, establish the true place of the formations
to which they belong. Like the patches of leather
of scarlet and of blue which mark the line attached
to the deep-sea lead, they show the various depths at
which we arrive. The Earls of Sutherland set themselves to establish a coal-work among the chambered
univalves of the Oolite, and a vast abundance of its
peculiar bivalves. The coal-borers who perforated the
Lias near Cromarty passed every day to and from
their work over one of the richest deposits of animal
remains in the kingdom,—a deposit full of the most
characteristic fossils ; and drove their augre through a
thousand belemnites and ammonites of the upper and
inferior Lias, and through gryphites and ichthyodorulites innumerable. The sandstones of Strathearn and
the Carse of Gowrie yield their plates and scales of
the *Holoptychius,* the most abundant fossil of the
Upper Old Red; and the shale of the little dell in
which the first Earl of Cromarty set his miners to
work, contains, as I have said, plates of the *Coccosteus* and scales of the *Osteolepis,*—fossils found only
in the Lower Old Red. Nature in all these localities
furnished the index, but men lacked the skill neces-

---

quadrant: navigational instrument, forerunner of the sextant; **deep-sea lead**: i.e. for
maritime navigation; a weighted line used to measure the depth of the sea for checking
one's position, the patches of material being traditional marks for particular depths;
near Cromarty: probably the Navity-Eathie area, or possibly north of the Sutors:
Torrens 2003.

sary to decipher it.*   I may mention that, independently of their well-marked organisms, there is a
simple test through which the lignites of the newer

* There occurs in Mr Murchison's *Silurian System* a singu-
larly-amusing account of one of the most unfortunate of all
coal-boring enterprizes ; the unlucky projector, a Welsh farmer, having set himself to dig for coal in the lowest member of
the system, at least six formations beneath the only one at
which the object of his search could have been found. Mr
Murchison thus relates the story :—

" At Tin-y-coed I found a credulous farmer ruining himself
in excavating a horizontal gallery in search of coal, an ignorant miner being his engineer. The case may serve as a striking example of the *coal-boring* mania in districts which cannot by possibility contain that mineral ; and a few words concerning it may, therefore, prove a salutary warning to those
who speculate for coal in the Silurian Rocks. The farm-house
of Tin-y-coed is situated on the sloping sides of a hill of trap,
which throw off, upon its north-western flank, thin beds of black
grauwacke shale, dipping to the west-north-west at a high angle.
The colour of the shale, and of the water that flowed down its
sides, the pyritous veins, and other vulgar symptoms of coal-
bearing strata, had long convinced the farmer that he possessed a large hidden mass of coal, and, unfortunately, a small
fragment of real anthracite was discovered, which burnt like
the best coal. Miners were sent for, and operations commenced. To sink a shaft was impracticable, both from the want
of means and the large volume of water. A slightly-inclined
gallery was therefore commenced, the mouth of which was
opened at the bottom of the hill, on the side of the little brook
which waters the dell. I have already stated, that in many
cases, where the intrusive trap throws off the shale, the latter
preserves its natural and unaltered condition to within a
certain distance of the trap ; and so it was at Tin-y-coed, for
the level proceeded for 155 feet with little or no obstacle.
Mounds of soft black shale attested the rapid progress of
the adventurers, when suddenly they came to a ' change of
mineral.' They were now approaching the nucleus of the little
ridge; and the rock they encountered was, as the men inform-

formations may be distinguished from the true coal of the carboniferous system. Coal, though ground into an impalpable powder, retains its deep black colour, and may be used as a black pigment; lignite, on the contrary, when fully levigated, assumes a reddish, or rather umbry hue.

I have said that the waters of the well of the coal-heugh are chalybeate,—a probable consequence of their infiltration through the iron oxides of the superior beds of the formation, and their subsequent passage through the deep red strata of the inferior bed. There could be very curious chapters written on mineral springs, in their connection with the formations through which they pass. Smollett's masterpiece, honest old Matthew Bramble, became thoroughly disgusted with the Bath waters on discovering that they filtered through an ancient burying-ground be-

ed me, ' *as hard as iron*,' viz. of lydianized schist, precisely analogous to that which is exposed naturally in ravines where all the phenomena are laid bare. The deluded people, however, endeavoured to penetrate the hardened mass, but the vast expense of blasting it put a stop to the undertaking, not, however, without a thorough conviction on the part of the farmer, that could he but have got through that hard stuff, he would most surely have been well recompensed, for it was just thereabouts that they began to find ' *small veins of coal*.' It has before been shown that portions of anthracite are not unfrequent in the altered shale, where it is in contact with the intrusive rock. And the occurrence of the smallest portion of anthracite is always sufficient to lead the Radnorshire farmer to suppose that he is very near ' El Dorado.' Amid all their failures I never met with an individual who was really disheartened ; a frequent exclamation being, ' Oh, if our squires were only men of *spirit*, we should have as fine coal as any in the world.' "—*Silurian System*, part i. p. 328.

Matthew Bramble: a character in the picaresque novel *The Expedition of Humphry Clinker* (1771) by Tobias Smollett.

longing to the Abbey, and that much of their peculiar taste and odour might probably be owing to the
" rotten bones and mouldering carcasses" through
which they were strained. Some of the springs of
the Old Red Sandstone have also the churchyard
taste, but the bones and carcasses through which they
strain are much older than those of the Abbey burying-ground at Bath. The bitumen of the strongly
impregnated rocks and clay-beds of this formation,
like the bitumen of the still more strongly impregnated limestones and shales of the Lias, seems to
have had rather an animal than vegetable origin.
The shales of the Eathic Lias burn like turf soaked
in oil, and yet they hardly contain one per cent. of
vegetable matter. In a single cubic inch, however,
I have counted about eighty molluscous organisms,
mostly ammonites, and minute striated scallops; and
the mass, when struck with the hammer, still yields
the heavy odour of animal matter in a state of decay.
The lower fish beds of the Old Red are, in some localities, scarcely less bituminous. The fossil scales
and plates which they enclose burn at the candle :
they contain small cavities filled with a strongly
scented semi-fluid bitumen, as adhesive as tar, and as
inflammable ; and for many square miles together the
bed is composed almost exclusively of a dark-coloured
semi-calcareous, semi-aluminous schist, scarcely less
fetid, from the great quantity of this substance which
it contains, than the swine-stones of England. Its
vegetable remains bear but a small proportion to its
animal organisms ; and from huge accumulations of
these last decomposing amid the mud of a still sea,

'**rotten bones ...**': quoted from one of Bramble's letters in *Humphry Clinker* (Miller has
added 'mouldering').

THE OLD RED SANDSTONE.

little disturbed by tempests or currents, and then suddenly interred by some widely-spread catastrophe, to ferment and consolidate under vast beds of sand and conglomerate, the bitumen* seems to have been elaborated. These bituminous schists, largely charged with sulphuret of iron, run far into the interior, along the flanks of the gigantic Ben Wevis, and through the exquisitely pastoral valley of Strathpeffer. The higher hills which rise over the valley are formed mostly of the great conglomerate,—Knockferril, with its vitrified fort,—the wooded and precipitous ridge over Brahan,—and the middle eminences of the gigantic mountain on the north; but the bottom and the lower slopes of the valley are occupied by the bituminous and sulphureous schists of the fish bed, and in these, largely impregnated with the peculiar ingredients of the formation, the famous medicinal springs of the Strath have their rise. They contain, as shown by chemical analysis, the sulphates of soda, of lime, of magnesia, common salt, and, above all, sulphureted hydrogen gas,—elements which masses of sea-mud charged with animal matter would yield as readily to the chemist as the medicinal springs of Strathpeffer. Is it not a curious reflection, that the commercial greatness of Britain in the present day should be closely connected with the towering and thickly-spread forests of arboraceous ferns and gigantic reeds —vegetables of strange form and uncouth names—

* "In the slaty schists of Seefeld, in the Tyrol," says Messrs Sedgwick and Murchison, " there is such an abundance of a similar bitumen, that it is largely extracted for medicinal purposes."—*Geol. Trans. for* 1829, p. 134.

Knockferril with its vitrified fort: the hill-fort on Knockfarril, like some others in the Highlands, had its stone walls partly fused by great heat from firing their wooden framework; whether deliberately or not is still debated; **analysis**: probably taken from Anderson 1834, p. 556.

which flourished and decayed on its surface age after age during the vastly-extended term of the carboniferous period, ere the mountains were yet upheaved, and when there was as yet no man to till the ground? Is it not a reflection equally curious, that the invalids of the present summer should be drinking health, amid the recesses of Strathpeffer, from the still more ancient mineral and animal debris of the lower ocean of the Old Red Sandstone, strangely elaborated for vast but unreckoned periods in the bowels of the earth? The fact may remind us of one of the specifics of a now obsolete school of medicine, which flourished in this country about two centuries ago, and which included in its *materia medica* portions of the human frame. Among these was the flesh of Egyptian mummies, impregnated with the embalming drugs,—the dried muscles and sinews of human creatures who had walked in the streets of Thebes or of Luxor three thousand years ago.

The commoner mineral springs of the formation, as might be anticipated from the very general diffusion of the oxide to which its owes its colour, are chalybeate. There are districts in Easter Ross and the Black Isle in which the traveller scarcely sees a runnel by the way-side that is not half-choked up by its fox-coloured coagulum of oxide. Two of the most strongly-impregnated chalybeates with which I am acquainted gush out of a sandstone-bed, a few yards apart, among the woods of Tarbat House, on the northern shore of the Frith of Cromarty. They splash among the pebbles with a half-gurgling, half-tinkling sound, in a solitary but not unpleasing recess, darken-

ed by alders and willows ; and their waters, after uniting in the same runnel, form a little melancholy-
looking *lochan*, matted over with weeds, and edged
with flags and rushes, and which swarms in early
summer with the young of the frog in its tadpole
state, and in the after months with the black water-
beetle and the newt. The circumstance is a somewhat
curious one, as the presence of iron as an oxide has
been held so unfavourable to both animal and vegetable life, that the supposed poverty of the Old Red
Sandstone in fossil remains has been attributed to its
almost universal diffusion at the period the deposition
was taking place. Were the system as poor as has
been alleged, however, it might be questioned, on the
strength of a fact such as this, whether the iron mili-
tated so much against the living existences of the formation as against the preservation of their remains
when dead.

Some of the springs which issue from the ichthyolite beds along the shores of the Moray Frith are
largely charged, not with iron, like the well of the
coal-heugh, or the springs of Tarbat House, nor yet
with hydrogen and soda, like the spa of Strathpeffer,
but with carbonate of lime. When employed for
domestic purposes, they choke up, in a few years,
with a stony deposition, the spouts of tea-kettles. On
a similar principle, they plug up their older channels,
and then burst out in new ones ; nor is it uncommon
to find among the cliffs little hollow recesses, long
since divested of their waters by this process, that are
still thickly surrounded by coral-like petrifactions of
moss and lichens, grass and nettle stalks, and roofed

---

lochan: small loch (Gaelic); **oxide**: possibly a misreading of Lyell (see pp. 21, 156).

with marble-like stalactites. I am acquainted with at least one of these springs, of very considerable volume, and dedicated of old to an obscure Roman Catholic saint, whose name it still bears (St Bennet), which presents phenomena not unworthy the attention of the young geologist. It comes gushing from out the ichthyolite bed, where the latter extends in the neighbourhood of Cromarty, along the shores of the Moray Frith; and after depositing in a stagnant morass an accumulation of a grayish-coloured and partially consolidated travertin, escapes by two openings to the shore, where it is absorbed among the sand and gravel. A storm about three years ago swept the beach several feet beneath its ordinary level, and two little moles of conglomerate and sandstone, the work of the spring, were found to occupy the two openings. Each had its fossils,—comminuted sea-shells, and stalks of hardened moss; and in one of the moles I found embedded a few of the vertebral joints of a sheep. It was a recent formation on a small scale, bound together by a calcareous cement furnished by the fish-beds of the inferior Old Red Sandstone, and composed of sand and pebbles, mostly from the granitic gneiss of the neighbouring hill, and organisms, vegetable and animal, from both the land and the sea.

The Old Red Sandstone of Scotland has been extensively employed for the purposes of the architect, and its limestones occasionally applied to those of the agriculturist. As might be anticipated in reference to a deposit so widely spread, the quality of both its sandstones and its lime is found to vary

---

St Bennet: the well flows onto the shore below the site of a chapel dedicated to the saint, probably originally named Bainan (Alston 2006, pp. 20, 26–27); moles: piers; architect: in the broader sense of a builder.

exceedingly in even the same beds when examined in different localities. Its inferior conglomerate, for instance, in the neighbourhood of Cromarty weathers so rapidly, that a fence built of stones furnished by it little more than half a century ago, has mouldered in some places into a mere grass-covered mound. The same bed in the neighbourhood of Inverness is composed of a stone nearly as hard and quite as durable as granite, and which has been employed in paving the streets of the place, —a purpose which it serves as well as any of the igneous or primary rocks could have done. At Redcastle, on the northern shore of the Frith of Beauly, the same conglomerate assumes an intermediate character, and forms, though coarse, an excellent building-stone, which, in some of the older ruins of the district, presents the marks of the tool as sharply indented as when under the hands of the workman. Some of the sandstone beds of the system are strongly saliferous; and these, however coherent they may appear, never resist the weather until first divested of their salt. The main ichthyolite bed on the northern shore of the Moray Frith is overlaid by a thick deposit of a finely-tinted yellow sandstone of this character, which, unlike most sandstones of a mouldering quality, resists the frosts and storms of winter, and wastes only when the weather becomes warm and dry. A few days of sunshine affect it more than whole months of high winds and showers. The heat crystallizes at the surface the salt which it contains; the crystals, acting as wedges, throw off minute particles of the stone; and thus, mechanically

N

---

overlaid: at Eathie Burn; see p. 211.

at least, the degrading process is the same as that to which sandstones of a different but equally inferior quality are exposed during severe frosts. In the course of years, however, this sandstone, when employed in building, loses its salt; crust after crust is formed on the surface, and either forced off by the crystals underneath or washed away by the rains; and then the stone ceases to waste, and gathers on its weathered inequalities a protecting mantle of lichens.* The most valuable quarries in the Old Red System of Scotland yet discovered are the flagstone quarries of Caithness and Carmylie. The former have been opened in the middle schists of the lower or Tilestone formation of the system; the latter in what has been deemed a cotemporaneous deposit, though, when the oryctology of the system comes to be better understood, it will, I doubt not, be demonstrated by its fossils to belong to the Cornstone or middle division. The quarries of both Carmylie and Caithness employ hundreds of workmen, and their flagstones form an article of commerce. The best building-stone of the north of Scotland,—best both for beauty

---

* When left to time the process is a tedious one, and, ere its accomplishment, the beauty of the masonry is always in some degree destroyed. The following passage, from a popular work, points out a mode by which it might possibly be anticipated, and the waste of surface prevented :—" A hall of which the walls were constantly damp, though every means were employed to keep them dry, was about to be pulled down, when M. Schmithall recommended, as a last resource, that the walls should be washed with sulphuric acid (vitriol). It was done, and the deliquescent salts being decomposed by the acid, the walls dried, and the hall was afterwards free from dampness." (*Recreations in Science.*)

---

schists: in the general sense of layered rock, rather than the technical term for a specific kind of metamorphic rock; *Recreations in Science*: an anonymous popular book of improving 'rational amusement' on assorted scientific topics: Anon. 1830, p. 115.

and durability,—is a pure Quartzose Sandstone furnished by the upper beds of the system. These are extensively quarried in Moray, near the village of Brughhead, and exported to all parts of the kingdom. The famous obelisk of Forres, so interesting to the antiquary,—which has been described by some writers as formed of a species of stone unknown in the district, and which, according to a popular tradition, was transported from the Continent, is evidently composed of this Quartzose Sandstone, and must have been dug out of one of the neighbouring quarries. And so coherent is its texture, that the storms of perhaps ten centuries have failed to obliterate its rude but impressive sculptures.

The limestones of both the upper and lower formations of the system have been wrought in Moray with tolerable success. In both, however, they contain a considerable per-centage of siliceous and argillaceous earth. The system, though occupying an intermediate place between two metalliferous deposits,—the grauwacke and the carboniferous limestone,—has not been found to contain workable veins anywhere in Britain, and in Scotland no metallic veins of any kind, with the exception of here and there a few slender threads of ironstone, and here and there a few detached crystals of galena. Its wealth consists exclusively in building and paving-stone, and in lime. Some of the richest tracts of corn-land in the kingdom rest on the Old Red Sandstone,—the agricultural valley of Strathmore, for instance, and the fertile plains of Easter-Ross ; Caithness has also its deep corn-bearing soils, and Moray

---

Brughhead: now Burghead; obelisk of Forres: Sueno's Stone, a 6.5-metre-high cross-slab and the largest Pictish stone of its kind still standing today (see note on p.161).

has been known for centuries as the granary of Scotland.   But in all these localities the fertility seems
derived rather from an intervening subsoil of tenacious diluvial clay, than from the rocks of the system.
Wherever the clay is wanting the soil is barren.   In
the moor of the Milbuy,—a tract about fifty square
miles in extent, and lying within an hour's walk of
the Friths of Cromarty and Beauly,—a thin covering
of soil rests on the sandstones of the lower formation.
And so extreme is the barrenness of this moor, that
notwithstanding the advantages of its semi-insular
situation, it was suffered to lie as an unclaimed common until about twenty-five years ago, when it was
parcelled out among the neighbouring proprietors.

Milbuy: Mealbuie, the crest of the hogback formed by the Black Isle.

physiognomy (p. 197): in the older sense of the science of deducing character from
facial structure, associated with the Swiss philosopher Johann Kaspar Lavater (1741–
1801). Miller's **idle or doubtful** indicates that, in his view, this science's claims were not
otherwise to be trusted; phrenology, a new science much debated in 1830s and 1840s
Edinburgh, used a similar approach to the structure of the human skull, and some of its
most important proponents (e.g. George Combe) were political reformers of a more

CHAPTER XI.

Geological Physiognomy.—Scenery of the Primary Formations; Gneiss, Mica Schist, Quartz Rock.—Of the Secondary; the Chalk Formations, the Oolite, the New Red Sandstone, the Coal Measures.—Scenery in the Neighbourhood of
Edinburgh.—Aspect of the Trap Rocks.—The Disturbing
and Denuding Agencies.—Distinctive Features of the Old
Red Sandstone.—Of the great Conglomerate.—Of the Ichthyolite Beds.—The Burn of Eathie.—The Upper Old Red
Sandstones.—Scene in Moray.

PHYSIOGNOMY is no idle or doubtful science in connection with Geology. The physiognomy of a country indicates almost invariably its geological character.
There is scarce a rock among the more ancient groupes
that does not affect its peculiar form of hill and valley.
Each has its style of landscape; and as the vegetation
of a district depends often on the nature of the underlying deposits, not only are the main outlines regulated
by the mineralogy of the formations which they define,
but also in many cases the manner in which these outlines are filled up. The colouring of the landscape is
well nigh as intimately connected with its geology as
the drawing. The traveller passes through a mountain-
ous region of gneiss. The hills, which, though bulky,

---

radical cast than Miller; **The traveller**: what follows on pp. 197–200 (up to the description of Edinburgh scenery on p. 200) is a sequence of idealised descriptions, some of
which are based on landscapes Miller had seen, others drawing on his reading and his
imagination; they do not necessarily represent real places, and at least one appears to be
a composite, which is why Miller (unusually for him) does not mention any place-
names.

are shapeless, raise their huge backs so high over the
brown dreary moors which, unvaried by precipice or
ravine, stretch away for miles from their feet, that
even amid the heats of midsummer the snow gleams in
streaks and patches from their summits. And yet so
vast is their extent of base, and their tops so truncated,
that they seem but half-finished hills notwithstanding,—hills interdicted somehow in the forming, and
the work stopped ere the upper storeys had been added.
He pursues his journey, and enters a district of micaceous schist. The hills are no longer truncated or the
moors unbroken ; the heavy ground-swell of the former
landscape has become a tempestuous sea, agitated by
powerful winds and conflicting tides. The picturesque and somewhat fantastic outline is composed of
high sharp peaks, bold craggy domes, steep broken
acclivities, and deeply-serrated ridges ; and the higher
hills seem as if set round with a frame-work of props
and buttresses, that stretch out on every side like the
roots of an ancient oak. He passes on, and the landscape varies : the surrounding hills, though lofty,
pyramidal, and abrupt, are less rugged than before ;
and the ravines, though still deep and narrow, are
walled by ridges no longer serrated and angular, but
comparatively rectilinear and smooth. But the vegetation is even more scanty than formerly ; the steeper
slopes are covered with streams of debris, on which
scarce a moss or lichen finds root; and the conoi-
dal hills, bare of soil from summit half-way to
base, seem so many naked skeletons, that speak of
the decay and death of nature. All is solitude and
sterility. The territory is one of Quartz Rock. Still

journey ... secondary districts (p. 199): the juxtaposition of **Quartz Rock** and
secondary (here, sandstone) rocks suggests that Miller was thinking of the trip from
Contin via Kinlochewe to Gairloch which he had made as a peripatetic stonemason in
1823 (Miller 1854a, pp. 239–49, 1993, pp. 236–47). The **busy population** reminds us that
Miller was lucky to see the Highlands before they were depopulated by clearance and
emigration, forced and otherwise – an important element in his later socio-political
writing (Miller 2022 [1858]; Taylor 2022a, pp. 29–31, 92–96, 2022b, pp. A25–A35).

the traveller passes on : the mountains sink into low swellings; long rectilinear ridges run out towards the distant sea, and terminate in bluff precipitous headlands. The valleys, soft and pastoral, widen into plains, or incline in long-drawn slopes of gentlest declivity. The streams, hitherto so headlong and broken, linger beside their banks, and then widen into friths and estuaries. The deep soil is covered by a thick mantle of vegetation,—by forest trees of largest growth, and rich fields of corn ; and the solitude of the mountains has given place to a busy population. He has left behind him the primary regions, and entered on one of the secondary districts.

And these less rugged formations have also their respective styles,—marred and obliterated often by the Plutonic agency, which imparts to them in some instances its own character, and in some an intermediate one, but in general distinctly marked and easily recognised. The Chalk presents its long inland lines of apparent coast, that send out their rounded headlands, cape beyond cape, into the wooded or corn-covered plains below. Here and there, there juts up at the base of the escarpment a white obelisk-like stack ; here and there, there opens into the interior a narrow grassy bay, in which noble beeches have cast anchor. There are valleys without streams ; and the scene atop is a scene of arid and uneven downs, that seem to rise and fall like the sea after a storm. We pass on to the Oolite : the slopes are more gentle, the lines of rising ground less continuous and less coast-like, the valleys have their rivulets, and the undulating surface is covered by a

---

apparent coast: i.e., eroded during a past deep marine submergence: Lyell 1838, pp. 334–35; **Oolite**: such as the Cotswolds and Lincolnshire Wolds.

richer vegetation. We enter on a district of New
Red Sandstone. Deep narrow ravines intersect elevated platforms. There are lines of low precipices so
perpendicular and so red that they seem as if walled
over with new brick; and here and there, amid the
speckled and mouldering sandstones that gather no
covering of lichen, there stands up a huge altar-like
mass of lime, mossy and gray, as if it represented a
remoter antiquity than the rocks around it. The
Coal Measures present often the appearance of vast
lakes, frozen over during a high wind, partially broken afterwards by a sudden thaw, and then frozen
again. Their shores stand up around them in the form
of ridges and mountain-chains of the older rocks; and
their surfaces are grooved into flat valleys and long
lines of elevation. Take as an instance the scenery
about Edinburgh. The Ochil Hills and the Grampians form the distant shores of the seeming lake or
basin on the one side, the range of the Lammermuirs
and the Pentland group on the other; the space between is ridged and furrowed in long lines, that run in
nearly the same direction from north-east to south-
west, as if, when the binding frost was first setting in,
the wind had blown from off the northern or southern
shore.

But whence these abrupt precipitous hills that stud
the landscape, and form, in the immediate neighbourhood of the city, its more striking features? They belong—to return to the illustration of the twice-frozen
lake—to the middle period of thaw, when the ice broke
up; and, as they are composed chiefly of matter eject-
ed from the abyss, might have characterized equally

New Red Sandstone: this strongly recalls Murchison's account of the western Midlands
of England, such as Hagley, Halesowen, and the vales of Severn and Teme, with the
mass of lime perhaps inspired by the outliers of Jurassic limestone capping isolated
hills, such as Bredon (Murchison 1839, part I, pp. 17–67).

any of the other formations.   Their very striking
forms, however, illustrate happily the operations of the
great agencies on which, in the secondary and transition deposits, all the peculiarities of scenery depend.
The molten matter from beneath seems to have been
injected, in the first instance, through rents and fissures among the carboniferous shales and sandstones
of the district, where it lay cooling in its subterranean
matrices, in beds and dikes, like metal in the moulds of
the founder; and the places which it occupied must
have been indicated on the surface but by curves and
swellings of the strata.   The denuding power then
came into operation in the form of tides and currents,
and ground down the superincumbent rocks.   The
injected masses, now cooled and hardened, were laid
bare ; and the softer framework of the moulds in which
they had been cast was washed from their summits
and sides, except where long ridges remained attached to them in the lines of the current, as if to indicate the direction in which they had broken its force.
Every larger stone in a water-course, after the torrent
fed by a thunder-shower has just subsided, shows, on
the same principle, its trail of sand and shingle piled
up behind it.   The outlines of the landscape were
modified yet farther by the yielding character of the
basement of sandstone or shale on which the Plutonic
beds so often rest.   The basement crumbled away as
the tides and waves broke against it.   The injected
beds above, undermined in the process, and with a
vertical cleavage, induced by their columnar tendency,
fell down in masses that left a front perpendicular
as a wall.   Each bed came thus to present its own

---

denuding power: in an 1810 study James Hall invoked a 'diluvian wave' (giant tsunami)
to explain the puzzling topography of the Edinburgh area, with its 'crag and tail'
structures such as the Castle Rock and Old Town (Hall 1814; Maclaren 1839; Rudwick
2005, pp.578–85). Other geologists such as Miller preferred a longer-lasting inundation
of the land by an iceberg-covered sea. But Agassiz suggested a solid glacial icecap
([Miller] 1840b; Finnegan 2004). See pp. A49–50, A63; **columnar**: like lava flows on
the surface (as on the isle of Staffa), flat sheets of intruded igneous rock, now called
sills, can suffer shrinkage cracks on cooling, dividing the rock into vertical columns.

upright line of precipice ; and hence—when they rise bed above bed, as often occurs—the stair-like outline of hill to which the trap-rocks owe their name ; hence the outline of the Dalmahoy Crags, for instance, and of the southern and western front of Salisbury Crags.

In all the sedimentary formations the peculiarities of scenery depend on three circumstances,—on the Plutonic agencies, the denuding agencies, and the manner and proportions in which the harder and softer beds of the deposits on which these operated, alternate with one another.   There is an union of the active and the passive in the formation of landscape ; that which disturbs and grinds down, and that which, according to its texture and composition, affects, if I may so speak, a peculiar style of being ground down and disturbed ; and it is in the passive circumstances that the peculiarities chiefly originate.   Hence it is that the scenery of the Chalk differs from the scenery of the Oolite, and both from that of the Coal Measures.   The Old Red Sandstone has also its peculiarities of prospect, which vary according to its formations, and the amount and character of the disturbing and denuding agencies to which these have been exposed.   Instead, however, of crowding its various, and, in some instances, dissimilar features into one landscape, I shall introduce to the reader a few of its more striking and characteristic scenes, as exhibited in various localities, and by different deposits, beginning first with its conglomerate base.

The great antiquity of this deposit is unequivocally indicated by the manner in which we find it caping,

name: from *trappa*, the Swedish for stair; see Glossary; **Dalmahoy Crags**: about 8 miles south-west of central Edinburgh; **Salisbury Crags**: on Arthur's Seat, the hill overlooking the Palace of Holyroodhouse in Edinburgh; **caping**: capping.

far in the interior, in insulated beds and patches, some of our loftier hills, or, in some instances, wrapping them round, as with a caul, from base to summit. It mixes largely, in our northern districts, with the mountain scenery of the country, and imparts strength and boldness of outline to every landscape in which it occurs. Its island-like patches affect generally a bluff parabolic or conical outline; its loftier hills present rounded dome-like summits, which sink to the plain on the one hand in steep, slightly concave lines, and on the other in lines decidedly convex, and a little less steep. The mountain of boldest outline in the line of the Caledonian Valley (Mealforvony), is composed externally of this rock. Except where covered by the diluvium, it seems little friendly to vegetation. Its higher summits are well nigh as bare as those of the primary rocks; and when a public road crosses its lower ridges, the traveller generally finds that there is no paving process necessary to procure a hardened surface, for that his wheels rattle over the pebbles embedded in the rock. On the sea coast, in several localities, the deposit presents striking peculiarities of outline. The bluff and rounded precipices stand out in vast masses, that affect the mural form, and present few of the minuter angularities of the primary rocks. Here and there a square buttress of huge proportions leans against the front of some low-browed crag, that seems little to need any such support, and casts a length of shadow athwart its face. There opens along the base of the rock a line of rounded shallow caves, or what seem rather

---

Mealforvony: Meall Fuar-mhonaidh, overlooking the north shore of Loch Ness; **Its higher summits**: from Sedgwick and Murchison 1829b, p. 129.

the openings of caves not yet dug, and which testify of a period when the sea stood about thirty feet higher on our coasts than at present. A multitude of stacks and tabular masses lie grouped in front, perforated often by squat heavy arches ; and stacks, caverns, buttresses, crags, and arches, are all alike mottled over by the thickly-set and variously-coloured pebbles. There is a tract of scenery of this strangely-marked character in the neighbourhood of Dunottar, and two other similar tracts in the far north, where the Hill of Nigg, in Ross-shire, declines towards the Lias deposit in the Bay of Shandwick, and where, in the vicinity of Inverness, a line of bold precipitous coast runs between the pyramidal wooded eminence which occupies the south-eastern corner of Ross, and the tower-like headlands that guard the entrance of the Bay of Munlochy. In the latter tract, however, the conglomerate is much less cavernous than in the other two.

We ascend to the middle and upper beds of this lower formation, and find scenery of a different character in the districts in which they prevail. The aspect is less bold and rugged, and affects often long horizontal lines, that stretch away, without rise or depression, amid the surrounding inequalities of the landscape, for miles and leagues, and that decline to either side, like roofs of what the architect would term a low pitch. The ridge of the Leys in the eastern opening of the Caledonian Valley, so rectilinear in its outline, and so sloping in its sides, presents a good illustration of this peculiarity. The rectilinear ridge

wooded eminence: Ord Hill, on the north coast of the Beauly Firth opposite Inverness. Miller was familiar with those coastlines from trips on the packet boats between Cromarty and Leith (for Edinburgh); **ridge**: about 5 miles south-east of Inverness, presumably that marked as Drummossie Muir on modern Ordnance Survey maps.

which runs from the Southern Sutor of Cromarty far
into the interior of the country, and which has been
compared in a former chapter to the shaft of a spear,
furnishes another illustration equally apt.* Where
the sloping sides of these roof-like ridges decline, as
in the latter instance, towards an exposed sea-coast,
we find the slope terminating often in an abrupt line
of rock dug out by the waves. It is thus a roof set
on walls and furnished with eaves. A ditch just
finished by the labourer presents regularly-sloping
sides; but the little stream that comes running through
gradually widens its bed by digging furrows into the
slopes, the undermined masses fall in and are swept
away, and in the course of a few months the sides are
no longer sloping, but abrupt. And such, on a great
scale, has been the process through which coast-lines
that were originally paved slopes have become walls
of precipices. The waves cut first through the outer
strata; and every stratum thus divided comes to present two faces,—a perpendicular face in the newly-
formed line of precipice, and another horizontal face

* The valleys which separate these ridges form often spaci-
ous friths and bays, the frequent occurrence of which in the
Old Red Sandstone constitutes, in some localities, one of the
characteristics of the system. Mark in a map of the north
of Scotland, how closely friths and estuaries lie crowded together between the counties of Sutherland and Inverness. In
a line of coast little more than forty miles in extent, there
occur four arms of the sea,—the Friths of Cromarty, Beauly,
and Dornoch, and the Bay of Munlochy. The Frith of Tay
and the Basin of Montrose are also semi-marine valleys of
the Old Red Sandstone. Two of the finest harbours in Britain,
or the world, belong to it,—Milford Haven in South Wales,
and the Bay of Cromarty.

Bay of Cromarty: the bay immediately west of the burgh, not the Firth of Cromarty as
a whole; not to be confused with the 'little bay' east of the town where Miller found his
fossil fishes (p. 113). Udale Bay is at the southern end of the Bay of Cromarty.

lying parallel to it, along the shore. One-half the se-
vered stratum seems as if rising out of the sea,—the
other half as if descending from the hill : the geologist who walks along the beach finds the various beds
presented in duplicate,—a hill-bed on the one side, and
a sea-bed on the other. There occurs a very interesting instance of this arrangement in the bold line of
coast on the northern shore of the Moray Frith, so often
alluded to in a previous chapter, as extending between
the Southern Sutor and the Hill of Eathie, and which
forms the wall of a portion of the roof-like ridge last
described. The sea first broke in a long line through
strata of red and gray shale, next through a thick bed
of pale-yellow stone, then through a continuous bed of
stratified clays and nodular limestone, and last of all
through a bed, thicker than any of the others, of indurated red sandstone. The line of cliffs formed in this
way rises abruptly for about a hundred yards on the
one hand ; the shore stretches out for more than
double the same space on the other ; on both sides the
beds exactly correspond ; and to ascend in the line of
the strata from the foot of the cliffs, we have either
to climb the hill, or to pass downwards at low ebb to
the edge of the sea. The section is of interest, not
only from the numerous organisms, animal and vegetable, which its ichthyolite beds contain, but from the
illustration which it also furnishes of denudation to a
vast extent from causes still in active operation. A
line of precipices a hundred yards in height, and more
than two miles in length, has been dug out of the
slope by the slow wear of the waves, in the unreckon
ed course of that period during which the present sea

was bounded in this locality by the existing line of coast. (See Frontispiece, section 3.)

I know not a more instructive walk for the young geologist than that furnished by the two miles of shore along which this section extends. Years of examination and inquiry would fail to exhaust it. It presents us, I have said, with the numerous organisms of the Lower Old Red Sandstone; it presents us also, towards its western extremity, with the still more numerous organisms of the Lower and Upper Lias: nor are the inflections and faults which its strata exhibit less instructive than its fossils or its vast denuded hollow. I have climbed along its wall of cliffs during the height of a tempestuous winter tide, when waves of huge volume, that had begun to gather strength under the night of the Northern Ocean, were bursting and foaming below; and as the harder pebbles, uplifted by the surge, rolled by thousands and tens of thousands along the rocky bottom, and the work of denudation went on, I have thought of the remote past, when the same agents had first begun to grind down the upper strata, whose broken edges now projected high over my head on the one hand, and lay buried far under the waves at my feet on the other. Almost all mountain chains present their abrupter escarpments to the sea, though separated from it in many instances by hundreds of miles,—a consequence, it is probable, of a similar course of denudation, ere they had attained their present altitude, or the plains at their feet had been elevated over the level of the ocean. Had a rise of a hundred feet taken place in this northern district in the days of Cæsar, the whole upper part of the Moray Frith would have been laid dry, and it would now have seemed as inexplicable that this roof-like ridge should present so rugged a line of wall to the distant sea, as that the Western Ghauts of India should invariably turn their steepest declivities to the basin of the Indian Ocean, or that from the arctic circle to the southern extremity of Patagonia the huge mountain-chain of America should elevate its dizzy precipices in the line of the Pacific.

Let us take another view of this section. It stretches between two of the granitic knobs or wedges to which I have had such frequent occasion to refer, —the Southern Sutor of Cromarty and the Hill of Eathie; and the edges of the strata somewhat remind one of the edges of a bundle of deals laid flatways on two stones, and bent towards the middle by their own weight. But their more brittle character is shown by the manner in which their ends are broken and uptilted against the granitic knobs on which they seem to rest; and towards the western knob the whole bundle has been broken across from below, and the opening occasioned by the fracture forms a deep savage ravine skirted by precipices, that runs far into the interior, and exhibits the lower portion of the system to well nigh its base. Will the reader spend a very few minutes in exploring the solitary recesses of this rocky trench,—it matters not whether as a scene-hunter or a geologist? We pass onwards along the beach through the middle line of the denuded hollow. The natural rampart that rises on the right ascends towards the uplands in steep slopes, lined horizontally by sheep walks, and fretted by

---

Western Ghauts: the Western Ghats or Sahyadri mountain range on the west coast of India; **deals**: planks; **ravine**: Eathie Den; **middle line of the denuded hollow**: i.e. along the line of the shore: compare pp. 205–7.

mossy knolls and churchyard-like ridges,—or juts out into abrupt and weathered crags, crusted with lichens and festooned with ivy,—or recedes into bosky hollows, roughened by the sloe-thorn, the wild-rose, and the juniper. On the left the wide extent of the Moray Frith stretches out to the dim horizon, with its vein-like currents, and its undulating lines of coast; while before us we see, far in the distance, the blue vista of the Great Valley, with its double wall of jagged and serrated hills, and, directly in the opening, the gray diminished spires of Inverness. We reach a brown mossy stream, of just volume enough to sweep away the pebbles and shells that have been strewed in its course by the last tide; and see, on turning a sudden angle, the precipices cleft to their base by the ravine, that has yielded its waters a passage from the interior.

We enter along the bed of the stream. A line of mural precipices rises on either hand,—here advancing in ponderous overhanging buttresses, there receding into deep damp recesses, tapestried with ivy, and darkened with birch and hazel. A powerful spring, charged with lime, comes pouring by a hundred different threads over the rounded brow of a beetling crag, and the decaying vegetation around it is hardening into stone. The cliffs vary their outline at every step, as if assuming in succession all the various combinations of form that constitute the wild and the picturesque; and the pale hues of the stone seem, when brightened by the sun, the very tints a painter would choose to heighten the effect of his shades, or to contrast most

o

---

mural: wall-like.

delicately with the luxuriant profusion of bushes and flowers that wave over the higher shelves and crannies. A colony of swallows have built from time immemorial under the overhanging strata of one of the loftier precipices ; the fox and badger harbour in the clefts of the steeper and more inaccessible banks. As we proceed, the dell becomes wilder and more deeply wooded ; the stream frets and toils at our feet,— here leaping over an opposing ridge,—there struggling in a pool,—yonder escaping to the light from under some broken fragment of cliff. There is a richer profusion of flowers, a thicker mantling of ivy and honey-suckle ; and, after passing a semicircular inflection of the bank, that waves from base to summit with birch, hazel, and hawthorn, we find the passage shut up by a perpendicular wall of rock about thirty feet in height, over which the stream precipitates itself in a slender column of foam, into a dark mossy basin. The long arms of an intermingled clump of birches and hazels stretch half-way across, tripling with their shade the apparent depth of the pool, and heightening in an equal ratio the white flicker of the cascade, and the effect of the bright patches of foam which, flung from the rock, incessantly revolve on the eddy.

Mark now the geology of the ravine. For about half-way from where it opens to the shore to where the path is obstructed by the deep mossy pool and the cascade, its precipitous sides consist of three bars or storeys. There is first, reckoning from the stream upwards, a broad bar of pale red,—then a broad bar of pale lead-colour,—last and highest, a broad bar of

pale yellow,—and above all there rises a steep green
slope, that continues its ascent till it gains the top of
the ridge. The middle lead-coloured bar is an ichthyolite bed, a place of sepulture among the rocks,
where the dead lie by myriads. The yellow bar
above is a thick bed of saliferous sandstone. We may
see the projections on which the sun has beat most
powerfully, covered with a white crust of salt ; and it
may be deemed worthy of remark, in connection with
the circumstance, that its shelves and crannies are
richer in vegetation than those of the other bars. The
pale-red bar below is composed of a coarser and harder
sandstone, which forms an upper moiety of the arenaceous portion of the great conglomerate. Now mark,
farther, that on reaching a midway point between the
beach and the cascade, this triple-barred line of precipices abruptly terminates, and a line of precipices of
coarse conglomerate as abruptly begins. I occasionally pass a continuous wall built at two different
periods, and composed of two different kinds of materials : the one half of it is formed of white sandstone,
—the other half of a dark-coloured basalt ; and the
place where the sandstone ends and the basalt begins
is marked by a vertical line, on the one side of which
all is dark coloured, while all is of a light colour on
the other. Equally marked and abrupt is the vertical line which separates the triple-barred from the
conglomerate cliffs of the ravine of Eathie. The ravine itself may be described as a fault in the strata ;
but here is a fault, lying at right angles with it, on a
much larger scale : the great conglomerate on which
the triple bars rest has been cast up at least two

hundred feet, and placed side by side with them.
And yet the surface above bears no trace of the catastrophe.   Denuding agencies of even greater power
than those which have hollowed out the cliffs of the
neighbouring coast, or whose operations have been
prolonged through periods of even more extended
duration, have ground down the projected line of the
upheaved mass to nearly the same level as that of the
undisturbed masses beside it.   Now, mark farther, as
we ascend the ravine, that the grand cause of the disturbance appears to illustrate, as it were, and that very
happily, the manner in which the fault was originally
produced.   The precipice over which the stream
leaps at one bound into the mossy hollow, is composed of granitic gneiss, and seems evidently to have
intruded itself, with much disturbance, among the
surrounding conglomerate and sandstones.   A few
hundred yards higher up the dell, there is another
much loftier precipice of gneiss, round which we find
the traces of still greater disturbance ; and higher
still, yet a third abrupt precipice of the same rock.
The gneiss rose, trap-like, in steps, and carried up
the sandstone before it in detached squares.   Each
step has its answering fault immediately over it ;
and the fault where the triple bars and the conglomerate meet is merely a fault whose step of granitic
gneiss stopped short ere it reached the surface.   But
the accompanying section (see Frontispiece, sect. 4)
will better illustrate the geology of this interesting
ravine, than it can be illustrated by any written description.   I may remark, ere taking leave of it, however, that its conglomerates exhibit a singularly large

amount of false stratification at an acute angle with
the planes of the real strata, and that a bed of mouldering sandstone near the base of the system may be
described, from its fissile character, as a tilestone.*

* There is a natural connection, it is said, beween wild
scenes and wild legends ; and some of the traditions connected
with this romantic and solitary dell illustrate the remark.
Till a comparatively late period it was known at many a winter fireside as a favourite haunt of the fairies,—the most poetical of all our old tribes of spectres, and at one time one of the
most popular. I have conversed with an old woman, who, when
a very little girl, had seen myriads of them dancing, as the sun
was setting, on the farther edge of the dell ; and with a still older
man, who had the temerity to offer one of them a pinch of snuff
at the foot of the cascade. Nearly a mile from where the ravine opens to the sea it assumes a gentler and more pastoral
character : the sides, no longer precipitous, descend towards
the stream in green sloping banks ; and a beaten path which
runs between Cromarty and Rosemarkie winds down the one
side, and ascends the other. More than sixty years ago, one
Donald Calder, a Cromarty shopkeeper, was journeying by
this path shortly after nightfall. The moon, at full, had just
risen ; but there was a silvery mist sleeping on the lower
grounds, that obscured her light ; and the dell, in all its extent,
was so overcharged by the vapour, that it seemed an immense
overflooded river winding through the landscape. Donald had
reached its farther edge, and could hear the rush of the stream
from the deep obscurity of the abyss below, when there rose
from the opposite side a strain of the most delightful music he
had ever heard. He staid and listened. The words of a song,
of such simple beauty that they seemed without effort to
stamp themselves on his memory, came wafted in the music ;
and the chorus, in which a thousand tiny voices seemed to join,
was a familiar address to himself,—" Ho, Donald Calder ; hey,
Donald Calder." " There are nane of my navity acquaintance," thought Donald, " who sing like that. Wha can it be ?"
He descended into the cloud ; but in passing the little stream
the music ceased ; and on reaching the spot on which the

false stratification: cross-bedding, indicating deposition by underwater currents running in one direction: perhaps Lyell 1838, pp. 38–40, and Murchison 1839, part I, p. 59;
I have conversed: Miller called the woman 'one of the perished volumes of my library'
of oral traditions, i.e. one of his deceased informants (Miller 1850b, p. 458; 1994, p. 442).
Among these may have been his grandparents Robert Wright (1733–1818) and Catherine
Rose (1738–1811), Catherine's sisters Isobel and Janet, and Janet's husband (Miller 1835,
pp. 2–3; Alston 1996; Marian McKenzie Johnston, pers. comm.). On these fairy stories,
see Additional Notes and pp. A78–81; **navity**: Navity, on the Cromarty side of Eathie.

I know comparatively little of the scenery of the middle or Cornstone formation. Its features in England are bold and striking,—in Scotland, of a tamer

singer had seemed stationed, he saw only a bare bank sinking into a solitary moor, unvaried by either bush or hollow in which the musician might have lain concealed. He had hardly time, however, to estimate the marvels of the case, when the music again struck up, but on the opposite side of the dell, and apparently from the very knoll on which he had so recently listened to it. The conviction that it could not be other than supernatural overpowered him; and he hurried homewards under the influence of a terror so extreme, that, unfortunately for our knowledge of fairy literature, it had the effect of obliterating from his memory every part of the song except the chorus. The sun rose as he reached Cromarty; and he found that, instead of having lingered at the edge of the dell for only a few minutes,—and the time had seemed no longer,—he had spent beside it the greater part of the night.

The fairies have deserted the Burn of Eathie; but we have proof quite as conclusive as the nature of the case admits, that when they ceased to be seen there it would have been vain to have looked for them anywhere else. There is a cluster of turf-built cottages grouped on the southern side of the ravine; a few scattered knolls and a long partially wooded hollow, that seems a sort of covered way leading to the recesses of the dell, interposes between them and the nearer edge; and the hill rises behind. On a Sabbath morning, nearly sixty years ago, the inmates of this little hamlet had all gone to church,—all except a herdboy and a little girl, his sister, who were lounging beside one of the cottages; when, just as the shadow of the garden-dial had fallen on the line of noon, they saw a long cavalcade ascending out of the ravine through the wooded hollow. It winded among the knolls and bushes; and, turning round the northern gable of the cottage, beside which the sole spectators of the scene were stationed, began to ascend the eminence towards the south. The horses were shaggy, diminutive things, spackled dun and gray; the riders, stunted, misgrown, ugly creatures, attired in antique

---

covered way: in fortifications of the time, a trench allowing the garrison to move to and fro while protected from observation and fire.

and more various character. The Den of Balruddery is a sweet wooded dell, marked by no characteristic peculiarities. Many of the seeming peculiarities of the formation in Forfarshire, as in Fife,
may be traced to the disturbing trap. The appearance exhibited is that of uneven plains, that rise and
fall in long ridges,—an appearance which any other
member of the system might have presented. We
find the upper formation associated with scenery of
great though often wild beauty; and nowhere is this
more strikingly the case than in the province of
Moray, where it leans against the granitic gneiss of
the uplands, and slopes towards the sea in long plains
of various fertility,—deep and rich, as in the neighbourhood of Elgin, or singularly bleak and unpro-
ductive, as in the far-famed " heath near Forres."
Let us select the scene where the Findhorn, after
hurrying over ridge and shallow amid combinations
of rock and wood, wildly picturesque as any the
kingdom affords, enters on the lower country, with a

jerkins of plaid, long gray cloaks, and little red caps, from
under which their wild uncombed locks shot out over their
cheeks and foreheads. The boy and his sister stood gazing in
utter dismay and astonishment, as rider after rider, each one
more uncouth and dwarfish than the one that had preceded it,
passed the cottage, and disappeared among the brushwood,
which at that period covered the hill, until at length the entire rout, except the last rider, who lingered a few yards behind the others, had gone by. " What are ye, little mannie?
and where are you going?" enquired the boy, his curiosity
getting the better of his fears and his prudence. " Not of the
race of Adam," said the creature, turning for a moment in its
saddle: " the People of Peace shall never more be seen in
Scotland."

---

heath near Forres: stage direction from Shakespeare's *Macbeth*, Act 1, Scene 3, referring
to the place where Macbeth and Banquo meet the three witches; '**Not of the race of
Adam**': in several Gaelic folktales, the suggestion that the fairies are not of Adam's race
is combined with the notion that they are fallen angels or their descendants, or angels
who had remained neutral during the legendary war between the good and bad angels:
Black 2005, pp. xxx–xxxi.

course less headlong, through a vast trench scooped in
the pale-red sandstone of the upper formation.   For
miles above the junction of the newer and older rocks
the river has been toiling in a narrow and uneven
channel, between two upright walls of hard gray
gneiss, thickly traversed, in every complexity of pattern, by veins of a light-red, large-grained granite.
The gneiss abruptly terminates, but not so the wall
of precipices.   A lofty front of gneiss is joined to a
lofty front of sandstone, like the front walls of two adjoining houses; and the broken and uptilted strata of
the softer stone show that the older and harder rocks
must have invaded it from below.   A little farther down the stream, the strata assume what seems,
in a short extent of frontage, a horizontal position,
like courses of ashler in a building, but which,
when viewed in the range, is found to incline at a
low angle towards the distant sea.   Here, as in many
other localities, the young geologist must guard
against the conclusion that the rock is necessarily low
in the geological scale, which he finds resting against
the gneiss.   The gneiss, occupying a very different
place from that on which it was originally formed,
has been thrust into close neighbourhood with widely-separated formations.   The great conglomerate base
of the system rests over it in Orkney, Caithness, Ross,
Cromarty, and Inverness; and there is no trace of
what should be the intervening grauwacke.   The
upper formation of the system leans upon it here.
We find the Lower Lias uptilted against it at the
Hill of Eathie,—the Great Oolite on the eastern
coast of Sutherland; and as the flints and chalk

---

lofty front of sandstone: this junction between the two formations is visible on the
banks of the Findhorn adjacent to the farm of Mains of Sluie; **ashler**: ashlar.

fossils of Banff and Aberdeen are found lying immediately over it in these counties, it is probable that
the denuded members of the Cretaceous group once
rested upon it there. The fact that a deposit should
be found lying in contact with the gneiss, furnishes
no argument for the great antiquity or the fundamental character of that deposit; and it were well
that the geologist who sets himself to estimate the
depth of the Old Red Sandstone, or the succession
of its various formations, should keep the circumstance in view. That may be in reality but a small
and upper portion of the system which he finds bounded by the gneiss on its under side, and by the diluvium on its upper.

We stand on a wooded eminence, that sinks perpendicularly into the river on the left, in a mural precipice, and descends with a billowy swell into the broad
fertile plain in front, as if the uplands were breaking
in one vast wave upon the low country. There is a
patch of meadow on the opposite side of the stream,
shaded by a group of ancient trees, gnarled and
mossy, and with half their topmost branches dead
and white as the bones of a skeleton. We look
down upon them from an elevation so commanding
that their uppermost twigs seem on well nigh the same
level with their interlaced and twisted roots, washed
bare on the bank edge by the winter floods. A colony
of herons has built from time immemorial among the
branches. There are trees so laden with nests that
the boughs bend earthwards on every side, like the
boughs of orchard-trees in autumn; and the bleached
and feathered masses which they bear,—the cradles of

---

wooded eminence: this description probably conflates more than one viewpoint on a
mile-long stretch of the Findhorn just to the north of Mains of Sluie; see Additional
Notes, Figure 14 and front cover of this volume.

succeeding generations,—glitter gray through the foliage in continuous groupes, as if each tree bore on its single head all the wigs of the Court of Session. The solitude is busy with the occupations and enjoyments of instinct. The birds, tall and stately, stand by troops in the shallows, or wade warily, as the fish glance by, to the edge of the current, or rising, with the slow flap of wing and sharp creak peculiar to the tribe, drop suddenly into their nests. The great forest of Darnaway stretches beyond, feathering a thousand knolls, that reflect a colder and grayer tint as they recede and lessen, and present on the horizon a billowy line of blue. The river brawls along under pale red cliffs, wooded atop. It is through a vast burial-yard that it has cut its way,—a field of the dead so ancient that the sepulchres of Thebes and Luxor are but of the present day in comparison,—resting-places for the recently departed, whose funerals are but just over. These mouldering strata are charged with remains, scattered and detached as those of a churchyard, but not less entire in their parts,—occipital bones, jaws, teeth, spines, scales,—the dust and rubbish of a departed creation. The cliffs sink as the plain flattens, and green sloping banks of diluvium take their place; but they again rise in the middle distance into an abrupt and lofty promontory, that, stretching like an immense rib athwart the level country, projects far into the stream, and gives an angular inflection to its course. There ascends from the apex a thin blue column of smoke,—that of a lime-kiln. That ridge and promontory are composed of the thick limestone band, which, in Moray as in Fife,

---

Court of Session: then the supreme civil court in Scotland, composed of the Lords of the Court of Session; Thebes and Luxor: the Thebes referred to here is an ancient Egyptian city, capital of Egypt during the New Kingdom period, whose ruins lie within the modern city of Luxor; immense rib: the hill with Limekilns and Cothall Woods north of Cothall on the Findhorn; the limestone worked hereabouts was more correctly a concretionary sandstone (Sedgwick and Murchison 1829b, pp. 151–52; Malcolmson 1859, pp. 341–42, 1921, pp. 447–52).

separates the pale-red from the pale-yellow beds of
the Upper Old Red Sandstone; and the flattened
tracts on both sides show how much better it has re-
sisted the denuding agencies than either the yellow
strata that rest over it, or the pale-red strata which
it overlies.

CHAPTER XII.

The two Aspects in which Matter can be viewed ; Space and Time.—Geological History of the Earlier Periods.—The Cambrian System.—Its Annelids.—The Silurian System.— Its Corals, Encrinites, Molluscs, and Trilobites.—Its Fish.— These of a high Order, and called into Existence apparently by Myriads.—Opening Scene in the History of the Old Red Sandstone a Scene of Tempest.—Represented by the Great Conglomerate.—Red a prevailing colour among the Ancient Rocks contained in this Deposit.—Amazing Abundance of Animal Life.—Exemplified by a Scene in the Herring Fishery.—Platform of Death.—Probable Cause of the Catastrophe which rendered it such.

" THERE are only two different aspects," says Dr Thomas Brown, " in which matter can be viewed. We may consider it simply as it exists in space, or as it exists in time. As it exists in space, we inquire into its composition, or, in other words, endeavour to discover what are the elementary bodies that co-exist in the space which it occupies ; as it exists in time, we inquire into its susceptibilities or its powers, or, in other words, endeavour to trace all the various changes which have already passed over it, or of which it may yet become the subject."

Hitherto I have very much restricted myself to the consideration of the Old Red Sandstone as it exists

Brown: abridged and paraphrased from a passage in the middle of 'On the Nature of Physical Inquiry in General', Lecture 5 in the extremely popular *Lectures on the Philosophy of the Human Mind* (1820) by the Scottish philosopher and physician Thomas Brown (1778–1820).

in *space*,—to the consideration of it as we now find it. I shall now attempt presenting it to the reader as it existed in *time*,—during the succeeding periods of its formation, and when its existences lived and moved as the denizens of primeval oceans. It is one thing to describe the appearance of a forsaken and desert country, with its wide wastes of unprofitable sand, its broken citadels and temples, its solitary battle-plains, and its gloomy streets of caverned and lonely sepulchres; and quite another to record its history during its days of smiling fields, populous cities, busy trade, and monarchical splendour. We pass from the dead to the living,—from the cemetery, with its high piles of mummies and its vast heaps of bones, to the ancient city, full of life and animation in all its streets and dwellings.

Two great geological periods have already come to their close; and the floor of a widely-spread ocean, to which we can affix no limits, and of whose shores or their inhabitants nothing is yet known, is occupied to the depth of many thousand feet by the remains of bygone existences. Of late the geologist has learned from Murchison to distinguish the rocks of these two periods,—the lower as those of the Cambrian, the upper as those of the Silurian group. The lower,—representative of the first glimmering twilight of being,—of a dawn so feeble that it may seem doubtful whether in reality the gloom had lightened,—must still be regarded as a period of uncertainty. Its ripple-marked sandstones, and its half-coherent accumulations of dark-coloured strata, which decompose into mud, show that

desert country: Egypt; **Murchison**: in fact he named the Cambrian jointly with Sedgwick at the British Association meeting of 1835 (Secord 1986, pp. 101–2).

every one of its many planes must have formed in succession an upper surface of the bottom of the sea ; but
it remains for future discoverers to determine regarding
the shapes of life that burrowed in its ooze, or career-
ed through the incumbent waters.   In one locality it
would seem as if a few worms had crawled to the surface, and left their involved and tortuous folds doubt-
fully impressed on the stone.   Some of them resemble
miniature cables carelessly coiled ; others, furnished
with what seem numerous legs, remind us of the existing Nereidina of our sandy shores,—those red-blooded,
many-legged worms, resembling elongated centipedes,
that wriggle with such activity among the mingled
mud and water, as we turn over the stones under
which they had sheltered.   Were creatures such as
these the lords of this lower ocean?   Did they enter
first on the stage, in that great drama of being in
which poets and philosophers, monarchs and mighty
conquerors, were afterwards to mingle as actors ?   Does
the reader remember that story in the *Arabian Nights*
in which the battle of the magicians is described ?
At an early stage of the combat a little worm creeps
over the pavement ; at its close two terrible dragons
contend in an atmosphere of fire.   But even the
worms of the Cambrian System can scarce be regarded
as established.   The evidence respecting their place
and their nature must still be held as involved in
some such degree of doubt as attaches to the researches
of the antiquary, when engaged in tracing what their
remains much resemble,—the involved sculpturings of
some Runic obelisk, weathered by the storms of a

one locality: near Llanbedr / Lampeter; evidently based on Murchison 1839, part I,
p. 363, part II, pp. 699–700 and plate 27; *Arabian Nights*: the story alluded to is the
'Second Dervish's Story' (Nights 13–14) in the frame-tale 'The Porter and the Three
Ladies', itself embedded in the *Thousand and One Nights* collection; the episode of two
shapechangers (a magically gifted princess and an *'ifrit*) is in Night 14: Lyons 2008, vol. I,
pp. 86–88; dragons: in the Arabic story these are sentient burning coals or flames, not
dragons.

thousand winters. There is less of doubt, however, regarding the existences of the upper group of rocks,—the Silurian.

The depth of this group, as estimated by Mr Murchison, is equal to double the height of our highest Scottish mountains ; and four distinct platforms of being range in it, the one over the other, like storeys in a building. Life abounded on all these platforms, and in shapes the most wonderful. The peculiar encrinites of the group rose in miniature forests, and spread forth their sentient petals by millions and tens of millions amid the waters ; vast ridges of corals, peopled by their innumerable builders,—numbers without number,—rose high amid the shallows ; the chambered shells had become abundant,—the simpler testacea still more so ; extinct forms of the graptolite or sea-pen existed by myriads ; and the formation had a class of creatures in advance of the many-legged annelids of the other. It had its numerous family of trilobites,—crustaceans nearly as high in the scale as the common crab,—creatures with crescent-shaped heads, and jointed bodies, and wonderfully-constructed eyes, which, like the eyes of the bee and the butterfly, had the cornea cut into facets resembling those of a multiplying-glass. Is the reader acquainted with the form of the common *Chiton* of our shores,—the little boat-shaped shell-fish that adheres to stones and rocks like the limpet, but which differs from every variety of limpet, in bearing as its covering a jointed, not a continuous shell? Suppose a chiton with two of its terminal joints cut away, and a single plate of much the same shape and size, but with two

---

four distinct platforms: i.e. levels or strata, rather than the physical kind of platform; in this case the Ludlow, Wenlock, Caradoc and Llandeilo formations made up the Silurian System in Murchison's classification (1839, part I, pp. 195–96); **Runic obelisk**: Pictish stele or cross-slab (see note on p. 161). Miller might have been thinking of Sueno's Stone, since he had used the same phrase **storms of a thousand winters** when describing that cross-slab on p. 195.

eyes near the centre, substituted instead, and the animal, in form at least, would be no longer a chiton, but
a trilobite. There are appearances, too, which lead to
the inference that the habits of the two families, though
representing different orders of being, may not have
been very unlike. The chiton attaches itself to the
rock by a muscular sucker or foot, which, extending
ventrally along its entire length, resembles that of
the slug or the snail, and enables it to crawl like
them, but still more slowly, by a succession of adhe-
sions. The locomotive powers of the trilobite seem to
have been little superior to those of the chiton. If
furnished with legs at all, it must have been with
soft rudimentary membranaceous legs, little fitted for
walking with ; and it seems quite as probable, from
the peculiarly-shaped under-margin of its shell, formed, like that of the chiton, for adhering to flat surfaces, that, like the slug and the snail, it was unfurnished with legs of any kind, and crept on the
abdomen. The vast conglomerations of trilobites for
which the Silurian rocks are remarkable are regarded
as further evidence of a sedentary condition. Like
Ostreæ, Chitones, and other sedentary animals, they
seem to have adhered together in vast clusters, trilobite over trilobite, in the hollows of submarine precipices, or on the flat muddy bottom below. And such
were the master existences of three of the four Silurian platforms, and of the greater part of the fourth,
if, indeed, we may not regard the chambered molluscs,
their cotemporaries,—creatures with their arms clustered round their heads, and with a nervous system composed of a mere knotted cord,—as equally high in the

---

chambered molluscs: shelled cephalopods; 'chambered shells' was a term he used for
these same creatures on p. 223.

scale. We rise to the topmost layers of the system,— to an upper gallery of its highest platform,—and find nature mightily in advance.

Another and superior order of existences had sprung into being at the fiat of the Creator,—creatures with the brain lodged in the head, and the spinal cord inclosed in a vertebrated column. In the period of the Upper Silurian, fish properly so called, and of very perfect organization, had become denizens of the watery element, and had taken precedence of the crustacean, as, at a period long previous, the crustacean had taken precedence of the annelid. In what form do these, the most ancient beings of their class, appear? As cartilaginous fishes of the higher order. Some of them were furnished with bony palates, and squat firmly-based teeth, well adapted for crushing the strong-cased zoophites and shells of the period, fragments of which occur in their fœcal remains ;—some with teeth that, like those of the fossil sharks of the later formations, resemble lines of miniature pyramids, larger and smaller alternating ;— some with teeth sharp, thin, and so deeply serrated that every individual tooth resembles a row of poniards set upright against the walls of an armoury; and these last, says Agassiz, furnished with weapons so murderous, must have been the pirates of the period. Some had their fins guarded with long spines, hooked like the beak of an eagle; some with spines of straighter and more slender form, and ribbed and furrowed longitudinally like columns; some were shielded by an armour of bony points; and some thickly covered with glistening scales. If many ages must have pass-

P

---

highest platform: i.e. the Ludlow, the highest formation of the Silurian; **fœcal**: error for 'fæcal'; **pirate**: perhaps inspired by the description by **Agassiz** as quoted in Murchison 1839, part II, p. 606.

ed ere fishes appeared, there was assuredly no time re-
quired to elevate their lower into their higher families.
Judging, too, from this ancient deposit, they seem to
have been introduced not by individuals and pairs,
but by whole myriads.

> " Forthwith the sounds and seas, each creek and bay,
>   With fry innumerable swarmed; and shoals
>   Of fish, that with their fins and shining scales
>   Glide under the green wave in plumps and sculls,
>   Banked the mid sea."

The fish-bed of the Upper Ludlow Rock abounds more
in osseous remains than an ancient burying-ground.
The stratum over wide areas seems an almost continuous layer of matted bones, jaws, teeth, spines, scales,
palatial plates, and shagreen-like prickles, all massed
together, and all converted into a substance of so deep
and shining a jet colour, that the bed, when " first
discovered, conveyed the impression," says Mr Murchison, " that it inclosed a triturated heap of black
beetles." And such are the remains of what seem to
have been the first existing vertebrata. Thus, ere our
history begins, the existences of two great systems, the
Cambrian and the Silurian, had passed into extinction, with the exception of what seem a few connecting links, exclusively molluscs, that are found in
England to pass from the higher beds of the Ludlow rocks into the lower or Tilestone beds of the Old
Red Sandstone. The exuviæ of at least four platforms of being lay entombed, furlong below furlong,
amid the gray mouldering mudstones, the harder
arenaceous beds, the consolidated clays, and the concretionary limestones, that underlay the ancient ocean

---

'**Forthwith ... sea**': adapted to Miller's tenses from Milton, *Paradise Lost*, book 7,
describing God's creation of the fishes and other water-dwellers; **fish-bed**: famous site
in Silurian rocks near Ludlow; **palatial**: error for 'palatal'; **The stratum**: Murchison
1839, part I, p.198, somewhat rewritten, and omitting Murchison's mention of
coprolites and his comments that although many fragments were jet black, others
were of a 'deep mahogany hue', the total effect being of 'a triturated heap of black
beetles cemented in a rusty ferruginous paste'.

of the Lower Old Red. The earth had already become a vast sepulchre, to a depth beneath the bed of the sea equal to at least twice the height of Ben Nevis over its surface.

The first scene in the *Tempest* opens amid the confusion and turmoil of the hurricane,—amid thunders and lightnings, the roar of the wind, the shouts of the seamen, the rattling of cordage, and the wild dash of the billows. The history of the period represented by the Old Red Sandstone seems, in what now forms the northern half of Scotland, to have opened in a similar manner. The finely-laminated lower Tilestones of England were deposited evidently in a calm sea. During the cotemporary period in our own country, the vast space which now includes Orkney and Lochness, Dingwall and Gamrie, and many a thousand square miles besides, was the scene of a shallow ocean, perplexed by powerful currents, and agitated by waves. A vast stratum of water-rolled pebbles, varying in depth from a hundred feet to a hundred yards, remains in a thousand different localities, to testify of the disturbing agencies of this time of commotion. The hardest masses which the stratum incloses,—porphyries of vitreous fracture that cut glass as readily as flint, and masses of quartz that strike fire quite as profusely from steel,—are yet polished and ground down into bullet-like forms, not an angular fragment appearing in some parts of the mass for yards together. The debris of our harder rocks, rolled for centuries in the beds of our more impetuous rivers, or tossed for ages along our more exposed and precipitous sea-shores, could not present less equivo-

---

Tempest: the play by Shakespeare; **cordage**: ship's rigging.

cally the marks of violent and prolonged attrition than the pebbles of this bed. And yet it is surely difficult to conceive how the bottom of any sea should have been so violently and so equally agitated for so greatly extended a space as that which intervenes between Mealforvonie in Inverness-shire and Pomona in Orkney in one direction, and between Applecross and Trouphead in another,—and for a period so prolonged, that the entire area should have come to be covered with a stratum of rolled pebbles of almost every variety of ancient rock, fifteen storeys height in thickness. The very variety of its contents shows that the period must have been prolonged. A sudden flood sweeps away with it the accumulated debris of a range of mountains; but to blend together, in equal mixture, the debris of many such ranges, as well as to grind down their roughnesses and angularities, and fill up the interstices with the sand and gravel produced in the process, must be a work of time. I have examined with much interest, in various localities, the fragments of ancient rock inclosed in this formation. Many of them are no longer to be found *in situ*, and the group is essentially different from that presented by the more modern gravels. On the shores of the Frith of Cromarty, for instance, by far the most abundant pebbles are of a blue schistose gneiss: fragments of gray granite and white quartz are also common; and the sea-shore at half-ebb presents at a short distance the appearance of a long belt of bluish gray, from the colour of the prevailing stones which compose it. The prevailing colour of the conglomerate of the district, on the contrary, is a

---

Mealforvonie: Meall Fuar-mhonaidh, see p. 203; **Pomona**: old name for Mainland, the principal island of Orkney.

deep red. It contains pebbles of small-grained red granite, red quartz rock, red feldspar, red porphyry, an impure red jasper, red hornstone, and a red granitic gneiss, identical with the well-marked gneiss of the neighbouring Sutors. This last is the only rock now found in the district, of which fragments occur in the conglomerate. It must have been exposed at the time to the action of the waves, though afterwards buried deep under succeeding formations, until again thrust to the surface by some great internal convulsion, of a date comparatively recent.*

The period of this shallow and stormy ocean passed. The bottom, composed of the identical conglomerate which now forms the summit of some of our loftiest mountains, sank throughout its wide area to a depth so profound as to be little affected by tides or

* The vast beds of unconsolidated gravel with which one of the later geological revolutions has half-filled some of our northern valleys, and covered the slopes of the adjacent hills, present, in a few localities, appearances somewhat analogous to those exhibited by this ancient formation. There are uncemented accumulations of water-rolled pebbles, in the neighbourhood of Inverness, from ninety to a hundred feet in thickness. But this stratum, unlike the more ancient one, wanted continuity. It must have been accumulated, too, under the operation of more partial, though immensely more powerful agencies. I do not remember having seen in the conglomerate of the Old Red a pebble which I could not have raised from the ground. There is a mediocrity of size in the enclosed fragments, which gives evidence of a mediocrity of power in the transporting agent. In the upper gravels, on the contrary, one of the agents could convey from vast distances blocks of stone eighty and a hundred tons in weight. A new cause of tremendous energy had come into operation in the geological world.

A new cause of tremendous energy: the powerful agent, of unknown but then much-debated nature which had shaped the surface of the country with such effect and left so much debris. Possibilities included a giant tsunami, iceberg-laden flood, or glacial ice-cap; see also note for p. 201.

tempests. During this second period there took place a vast deposit of coarse sandstone strata, with here and there a few thin beds of rolled pebbles. The general subsidence of the bottom still continued, and after a deposit of fully ninety feet had overlain the conglomerate, the depth became still more profound than at first. A fine semi-calcareous, semi-aluminous deposition took place in waters perfectly undisturbed. And here we first find proof that this ancient ocean literally swarmed with life,—that its bottom was covered with miniature forests of algæ, and its waters darkened by immense shoals of fish.

In middle autumn, at the close of the herring season, when the fish have just spawned, and the congregated masses are breaking up on shallow and skerry, and dispersing by myriads over the deeper seas, they rise at times to the surface by a movement so simultaneous, that for miles and miles around the skiff of the fisherman nothing may be seen but the bright glitter of scales, as if the entire face of the deep were a blue robe spangled with silver. I have watched them at sunrise at such seasons on the middle of the Moray Frith, when, far as the eye could reach, the surface has been ruffled by the splash of fins, as if a light breeze swept over it, and the red light has flashed in gleams of an instant on the millions and tens of millions that were leaping around me, a handbreadth into the air, thick as hailstones in a thunder-shower. The amazing amount of life which the scene included has imparted to it an indescribable interest. On most occasions the inhabitants of ocean are seen but by scores and hundreds ; for in looking

---

herring season: one of Miller's earliest literary successes was a series of letters on the Moray Firth herring fishery, published in the *Inverness Courier* in summer 1829, and then reissued as a pamphlet. This passage is a revised, expanded version of an anecdote in it, reusing elements such as the blue robe and the light breeze, and enlarging its invocation of divine plenitude: [Miller] 1829, pp. 36–37.

down into their green twilight haunts, we find the view bounded by a few yards, or at most a few fathoms; and we can but calculate on the unseen myriads of the surrounding expanse, by the seen few that occupy the narrow space visible. Here, however, it was not the few, but the myriads, that were seen,—the innumerable and inconceivable whole,—all palpable to the sight as a flock on a hill-side; or at least, if all was not palpable, it was only because sense has its limits in the lighter as well as in the denser medium,—that the multitudinous distracts it, and the distant eludes it, and the far horizon bounds it. If the scene spoke not of infinity in the sense in which Deity comprehends it, it spoke of it in at least the only sense in which man can comprehend it.

Now, we are much in the habit of thinking of such amazing multiplicity of being,—when we think of it at all,—with reference to but the later times of the world's history. We think of the remote past as a time of comparative solitude. We forget that the now uninhabited desert was once a populous city. Is the reader prepared to realize, in connection with the Lower Old Red Sandstone,—the second period of vertebrated existence,—scenes as amazingly fertile in life as the scene just described,—oceans as thoroughly occupied with being as our friths and estuaries when the herrings congregate most abundantly on our coasts? There are evidences too sure to be disputed that such must have been the case. I have seen the ichthyolite beds, where washed bare in the line of the strata, as thickly covered with oblong spindle-shaped nodules as I have ever seen a fishing-bank covered with herrings; and have ascertained that every individual nodule had its nucleus of animal matter,—that it was a stone coffin in miniature, holding inclosed its organic mass of bitumen or bone,—its winged, or enamelled, or thorn-covered ichthyolite.

At this period of our history some terrible catastrophe involved in sudden destruction the fish of an area at least a hundred miles from boundary to boundary, perhaps much more. The same platform in Orkney as at Cromarty is strewed thick with remains, which exhibit unequivocally the marks of violent death. The figures are contorted, contracted, curved; the tail in many instances is bent round to the head; the spines stick out; the fins are spread to the full, as in fish that die in convulsions. The *Pterichthys* shows its arms extended at their stiffest angle, as if prepared for an enemy. The attitudes of all the ichthyolites on this platform are attitudes of fear, anger, and pain. The remains, too, appear to have suffered nothing from the after attacks of predaceous fishes: none such seem to have survived. The record is one of destruction at once widely spread, and total so far as it extended. There are proofs, that whatever may have been the cause of the catastrophe, it must have taken place in a sea unusually still. The scales, when scattered by some slight undulation, are scattered to the distance of only a few inches, and still exhibit their enamel entire, and their peculiar fineness of edge. The spines, even when separated, retain their original needle-like sharpness of point. Rays well nigh as slender as horse hairs are inclosed unbroken in the mass. Whole ichthyolites occur, in which not only all

---

platform: not a topographic platform, but Miller's word for what might be called a geological horizon today. He is claiming that the Orcadian and Cromarty fish beds are at the same stratigraphic horizon, and therefore contemporaneous, and interprets them as remains of the same mass-mortality event.

the parts survive, but even the expression which the
stiff and threatening attitude conveyed when the last
struggle was over.   Destruction must have come in the
calm, and it must have been of a kind by which the
calm was nothing disturbed.   In what could it have
originated ?   By what quiet but potent agency of
destruction were the innumerable existences of an
area perhaps ten thousand square miles in extent annihilated at once, and yet the medium in which they
had lived left undisturbed by its operations ?   Conjecture lacks footing in grappling with the enigma, and
expatiates in uncertainty over all the known phenomena of death.   Diseases of mysterious origin break
out at times in the animal kingdom, and well nigh
exterminate the tribes on which they fall.   The present generation has seen a hundred millions of the human family swept away by a disease unknown to our
fathers.   Virgil describes the fatal murrain that once
depopulated the Alps, not more as a poet than as a historian.   The shell-fish of the rivers of North America
died in such vast abundance during a year of the present century, that the animals, washed out of their
shells, lay rotting in masses beside the banks, infecting
the very air.   About the close of the last century the
haddock well nigh disappeared for several seasons
together from the eastern coasts of Scotland ; and it is
related by Creech, that a Scotch shipmaster of the
period sailed for several leagues on the coast of Norway, about the time the scarcity began, through a
floating shoal of dead haddocks.*   But the ravages of

* I have heard elderly fishermen of the Moray Frith state,
in connection with what they used to term " the haddock

disease: the Second Pandemic of Asiatic cholera, or 'cholera morbus', hit Britain in 1831,
and Miller later recalled its impact on Cromarty (Miller 1854a, pp. 461–69, 1993, pp. 459–
67). It was not fully recognised as a disease distinct from other 'choleras', i.e. severe
diarrhoeas in general, till the First Pandemic of 1816–26, and was therefore **unknown
to our fathers**; **murrain**: disease of farm animals, described in the third book of the
Georgics (29 BC) by the Roman poet Virgil; Miller presumably used Dryden's translation;
Creech: William Creech (1745–1815), publisher of Robert Burns, paraphrased (and
quoted on p. 234) from Creech 1793, p. 51.

no such disease, however extensive, could well account for some of the phenomena of this platform of death. It is rarely that disease falls equally on many different

dearth" of this period, that for several weeks ere the fish entirely disappeared, they acquired an extremely disagreeable taste, as if they had been boiled in tobacco-juice, and became unfit for the table. For the three following years they were extremely rare on the coast, and several years more elapsed ere they were caught in the usual abundance. The fact related by Creech, a very curious one, I subjoin in his own words: it occurs in his third *Letter to Sir John Sinclair* :— " On Friday the 4th of December 1789, the ship Brothers, Captain Stewart, arrived at Leith from Archangel, who reported that, on the coast of Lapland and Norway, he sailed many leagues through immense quantities of dead haddocks floating on the sea. He spoke several English ships, who reported the same fact. It is certain that haddocks, which was the fish in the greatest abundance in the Edinburgh market, has scarcely been seen there these three years.   In February 1790 three haddocks were brought to market, which, from their scarcity, sold for 7s. 6d."

The dead haddocks seen by the Leith shipmaster were floating by thousands ; and most of their congeners among what fishermen term " the white fish," such as cod, ling, and whiting, also float when dead ; whereas the bodies of fish whose bowels and air-bladders are comparatively small and tender, lie at the bottom. The herring-fisherman, if the fish die in his nets, finds it no easy matter to buoy them up ; and if the shoal entangled be a large one, he fails at times, from the great weight, in recovering them at all, losing both nets and herrings. Now if a corresponding difference obtained among fish of the extinct period,—if some rose to the surface when they died, while others remained at the bottom,—we must of course expect to find their remains in very different degrees of preservation,—to find only scattered fragments of the floaters, while of the others many may occur comparatively entire. Even should they have died on the same beds, too, we may discover their remains separated by hundreds of miles. The haddocks that disappeared from the coast of Britain

tribes at once, and never does it fall with instantaneous suddenness; whereas in the ruin of this platform
from ten to twelve distinct genera seem to have been
equally involved; and so suddenly did it perform its
work, that its victims were fixed in their first attitude
of terror and surprise. I have observed, too, that
groupes of adjoining nodules are charged frequently
with fragments of the same variety of ichthyolite;
and the circumstance seems fraught with evidence regarding both the original habits of the creatures, and
the instantaneous suddenness of the destruction by
which they were overtaken. They seem, like many
of our existing fish, to have been gregarious, and to
have perished together ere their crowds had time to
break up and disperse.

Fish have been found floating dead in shoals beside submarine volcanoes,—killed either by the heated
water or by mephitic gases. There are, however, no
marks of volcanic activity in connection with the
ichthyolite beds,—no marks at least which belong to
nearly the same age with the fossils. The disturbing

were found floating in shoals on the coasts of Norway. The
remains of an immense body of herrings that weighed down,
a few seasons since, the nets of a crew of fishermen, in a muddy hollow of the Moray Frith, and defied the utmost exertions
of three crews united to weigh them from the bottom, are, I
doubt not, in the muddy hollow still. On a principle thus obvious it may be deemed not improbable that the ichthyolites
of the Lower Old Red Sandstone might have had numerous
cotemporaries, of which, unless in some instances the same
accident which killed also entombed them, we can know nothing in their character as such, and whose broken fragments
may yet be found in some other locality, where they may be
regarded as characteristic of a different formation.

---

nets: Miller may be conflating similar, but not identical, incidents from his *Letters from the Herring Fishing* (Miller] 1829, pp.16–17, 29–30), or simply reporting a more recent event; see also Miller 1859, pp.244–46.

granite of the neighbouring eminences was not upheaved until after the times of the Oolite. But the
volcano, if such was the destroying agent, might have
been distant; nay, from some of the points in an
area of such immense extent, it *must* have been distant. The beds abound, as has been said, in lime;
and the thought has often struck me that calcined
lime, cast out as ashes from some distant crater, and
carried by the winds, might have been the cause of
the widely-spread destruction to which their organisms testify. I have seen the fish of a small trouting
stream, over which a bridge was in the course of
building, destroyed in a single hour, for a full mile
below the erection, by the few troughfuls of lime that
fell into the water when the centring was removed.

CHAPTER XIII.

Successors of the exterminated Tribes.—The Gap slowly filled.—Proof that the Vegetation of a Formation may long
survive its Animal Tribes.—Probable Cause.—Immensely
extended Period during which Fishes were the Master
Existences of our Planet.—Extreme Folly of an Infidel Objection illustrated by the Fact.—Singular Analogy between
the History of Fishes as Individuals and as a Class.—Chemistry of the Lower Formation.—Principles on which the
Fish-inclosing Nodules were probably formed.—Chemical
Effect of Animal Matter in discharging the Colour from Red
Sandstone.—Origin of the prevailing Tint to which the
System owes its Name.—Successive Modes in which a Metal
may exist.—The Restorations of the Geologist void of Colour.—Very different Appearance of the Ichthyolites of Cromarty and Moray.

THE period of death passed, and over the innumerable dead there settled a soft muddy sediment, that
hid them up from the light, bestowing upon them such
burial as a November snow-storm bestows on the sere
and blighted vegetation of the previous summer and
autumn.   For an unknown space of time, represented
in the formation by a deposit about fifty feet in thickness, the waters of the depopulated area seem to have
remained devoid of animal life.   A few scales and
plates then begin to appear.   The fish that had existed

outside the chasm seem to have gradually gained upon it, as their numbers increased, just as the European settlers of America have been gaining on the backwoods, and making themselves homes amid the burial mounds of a race extinct for centuries. For a lengthened period, however, these finny settlers must have been comparatively few,—mere squatters in the waste. In the beds of stratified clay in which their remains first occur, over what we may term the densely-crowded platform of violent death, the explorer may labour for hours together without finding a single scale.

It is worthy of remark, however, that this upper bed abounds quite as much in the peculiar vegetable impressions of the formation as the lower platform itself. An abundance equally great occurs in some localities only a few inches over the line of the exterminating catastrophe. Thickets of exactly the same algæ amid which the fish of the formation had sheltered when living, grew luxuriantly over their graves when dead. The agencies of destruction which annihilated the animal life of so extended an area, spared its vegetation; just as the identical forests that had waved over the semi-civilized aborigines of North America continued to wave over the more savage red men, their successors, long after the original race had been exterminated. The inference deducible from the fact, though sufficiently simple, seems in a geological point of view a not unimportant one. *The flora of a system may long survive its fauna ; so that that may be but one formation, regarded with reference to plants, which may be two or more formations, regarded with reference to animals.* No instance of any such phenomenon occurs

---

red men: a once-standard, now-pejorative term for various indigenous North American peoples. Around 1840 it was commonly believed that the ancient earthworks of the Midwest were built by an earlier civilisation which the current Native Americans had superseded, just as Germanic peoples had conquered ancient Rome. This view was repeated in Cherokee legends and widely circulated in William Bartram's *Travels* (1791). Miller here uses **exterminated** in its non-purposive biogeographical sense, meaning 'caused the extinction of' rather than 'murdered'.

in the later geological periods. The changes in animal and vegetable life appear to have run parallel to
each other from the times of the tertiary formations
down to those of the coal ; but in the earlier deposits
the case must have been different.   The animal organisms of the newer Silurian strata form essentially different groupes from those of the Lower Old Red Sandstone, and both differ from those of the Cornstone divisions ; and yet the greater portion of their vegetable remains seem the same. The stem-like impressions of the
fucoid bed of the Upper Ludlow Rocks cannot be distinguished from those of the ichthyolite beds of Cromarty and Ross, nor these again from the impressions
of the Arbroath pavement or the Den of Balruddery.
Nor is there much difficulty in conceiving how the vegetation of a formation should come to survive its animals.
What is fraught with health to the existences of the vegetable kingdom is in many instances a deadly poison
to those of the animal.   The grasses and water-lilies of
the neighbourhood of Naples flourish luxuriantly amid
the carbonic acid gas which rests so densely over the
pools and runnels out of which they spring, that the
bird stoops to drink, and falls dead into the water.
The lime that destroys the reptiles, fish, and insects,
of a thickly-inhabited lake or stream, injures not a
single flag or bulrush among the millions that line
its edges.   The two kingdoms exist under laws of life
and death so essentially dissimilar, that it has become
one of the common-places of poetry to indicate the
blight and decline of the tribes of the one by the un-
wonted luxuriancy of the productions of the other.
Otway tells us, in describing the horrors of the plague

---

reptiles: including amphibians; **Otway**: Thomas Otway (1652–85), English Restoration
playwright and poet; **plague**: the 'Great Plague of London', an outbreak of bubonic
plague which hit southern and eastern England in 1665–66. It is remembered (as here)
for devastating London, but several other areas lost a greater proportion of their populations.

which almost depopulated London, that the " destroying angel stretched his arm" over the city,

> " Till in th' untrodden streets unwholesome grass
> Grew of great stalk, and colour gross,
> A melancholic poisonous green."

The work of deposition went on : a bed of pale yellow saliferous sandstone settled, tier over tier, on a bed
of stratified clay, and was itself overlaid by another bed
of stratified clay in turn.   And this upper bed had also
its organisms.   The remains of its sea-weed still spread
out thick and dark amid the foldings of the strata,
and occasionally its clusters of detached scales.   But
the circumstances were less favourable to the preservation of entire ichthyolites than those under which
the organisms of the lower platform were wrapped up
in their stony coverings.   The matrix, which is more
micaceous than the other, seems to have been less
conservative, and the waters were probably less still.
Data is wanting, in the obscurity of the remains, to
decide whether a change of species took place in consequence of the catastrophe of the lower platform.
Nothing, however, is more probable. The process went
on.   Age succeeded age, and one stratum covered up
another.   Generations lived, died, and were entombed
in the ever-growing depositions.   Succeeding generations pursued their instincts by myriads, happy in
existence, over the surface which covered the broken
and perishing remains of their predecessors, and then
died and were entombed in turn, leaving a higher
platform and a similar destiny to the generations that
succeeded. Whole races became extinct, through
what process of destruction who can tell? Other races

'destroying angel ... arm': paraphrased from Otway's satirical poem *The Poet's Complaint of His Muse* (1680), using the lines immediately preceding the quoted passage, in which 'a destroying angel was sent down ... and o'er all Britain stretch'd his conquering hand' to inflict the plague (see also 2 Samuel 24:16 in the Bible for the source of the 'destroying angel' image); 'Till in ... green': from Otway's *Poet's Complaint*; **change of species**: see Additional Notes.

sprang into existence through that adorable power which One only can conceive, and One only can exert. An inexhaustible variety of design expatiated freely within the limits of the ancient type. The main conditions remained the same,—the minor details were dissimilar. Vast periods passed; a class low in the scale still continued to furnish the master-existences of creation; and so immensely extended was the term of its sovereignty, that a being of limited faculties, if such could have existed uncreated, and witnessed the whole, would have inferred that the power of the Creator had reached its extreme boundary when fishes had been called into existence, and that our planet was destined to be the dwelling-place of no nobler inhabitants. If there be men dignified by the name of philosophers who can hold that the present state of being, with all its moral evil and all its physical suffering, is to be succeeded by no better and happier state, just because " all things have continued as they were" for some five or six thousand years, how much sounder and more conclusive would the inference have been which could have been based, as in the supposed case, on a period perhaps a hundred times more extended.

There exist wonderful analogies in nature between the geological history of the vertebrated animals as an order, and the individual history of every mammifer, —between the history, too, of fish as a class, and that of every single fish. " It has been found by Tiedemann," says Mr Lyell, " that the brain of the fœtus in the higher class of vertebrated animals assumes in succession the various forms which belong to fishes, reptiles, and birds, before it acquires those additions

Q

---

and modifications which are peculiar to the mammiferous tribes." " In examining the brain of the mammalia," says M. Serres, " at an early stage of life, you perceive the cerebral hemispheres consolidated, as in fish, in two vesicles isolated one from the other; at a later period you see them affect the configuration of the cerebral hemispheres of reptiles; still later, again, they present you with the forms of those of birds; and finally, at the era of birth, the permanent forms which the adult mammalia present." And such seems to have been the history of the vertebrata as an order, as certainly as that of the individual mammifer. The fish preceded the reptile in the order of creation, just as the crustacean had preceded the fish, and the annelid the crustacean. Again, the reptile seems to have preceded the bird. We find, however, unequivocal traces of the feathered tribes in well marked foot-prints impressed on a sandstone in North America, at most not more modern than the Lias, but which is generally supposed to be of the same age with the New Red Sandstone of Germany and our own country. In the Oolite,—at least one, perhaps two formations later,—the bones of the two species of mammiferous quadrupeds have been found, apparently of the marsupial family; and these, says Mr Lyell, afford the only example yet known of terrestrial mammalia in rocks of a date anterior to the older tertiary formations. The reptile seems to have preceded the bird, and the bird the mammiferous animal. Thus the fœtal history of the nervous system in the individual mammifer seems typical, in every stage of its progress, of the history of the grand division at the

Serres: Étienne Serres (1786–1868), French anatomist and embryologist. Miller is here quoting the paraphrase in Lyell 1830–33, vol. II, p. 63; **foot-prints**: in New Jersey, later reinterpreted as made by dinosaurs, themselves recognised as a distinct group of reptiles only in April 1842 (Torrens 2012); **Oolite … mammiferous quadrupeds**: from Lyell 1830–33, vol. I, p. 150 – though Lyell was even more sceptical than Miller about the notion of linear progress in the fossil record.

Agassiz (p. 243): probably from press reports of Agassiz's paper such as Anon. 1840i and

head of which the mammifer stands. Agassiz, at
the late meeting of the British Association in Glasgow, mentioned an analogous fact. After describing
the one-sided tail of the more ancient fish, especially
the fish of the Old Red Sandstone,—the subjects of
his illustration at the time,—he stated, as the result of
a recent discovery, that the young of the salmon in
their fœtal state exhibit the same unequally-sided condition of tail which characterises those existences of
the earlier ages of the world. The individual fish,
just as it begins to exist, presents the identical appearances which were exhibited by the order when the
order began to exist. Is there nothing wonderful in
analogies such as these,—analogies that point through
the embryos of the present time to the womb of Nature, big with its multitudinous forms of being? Are
they charged with no such nice evidence as a Butler
would delight to contemplate, regarding that unique
style of Deity, if I may so express myself, which runs
through all his works, whether we consider him as
God of Nature, or Author of Revelation ! In this
style of type and symbol did He reveal himself of old
to his chosen people : in this style of allegory and
parable did he again address himself to them, when
he sojourned among them on earth.

The chemistry of the formation seems scarce inferior
in interest to its zoology ; but the chemist has still
much to do for Geology, and the processes are but imperfectly known. There is no field in which more
laurels await the philosophical chemist than the geological one. I have said that all the calcareous nodules of the ichthyolite beds seem to have had origi-

---

1840m; his salmon work was later published in abstract as Agassiz 1841; **womb of
Nature**: in Milton's *Paradise Lost*, book 2, line 911, Satan uses this phrase to refer to the
primordial Chaos from which some early modern philosophers believed the world had
emerged at Creation (Forsyth 2009); but Miller seems to use the phrase purely metaphorically (compare Pidgeon's phrasing quoted in the Additional Note to p. 274); **nice**:
delicate, subtle; **Butler**: see Additional Notes; **reveal himself**: to the ancient Israelites,
as seen in the Old Testament; **sojourned among them**: i.e. as Jesus Christ; the phrase
is repeatedly used in the Bible of strangers or temporary visitors, including Jesus.

nally their nucleus of organic matter. In nine cases
out of ten the organism can be distinctly traced; and
in the tenth there is almost always something to indicate where it lay,—an elliptical patch of black, or an
oblong spot, from which the prevailing colour of the
stone has been discharged, and a lighter hue substituted. Is the reader acquainted with Mr Pepys' ac-
cidental experiment, as related by Mr Lyell, and recorded in the first volume of the *Geological Transactions*? It affords an interesting proof that animal
matter, in a state of putrefaction, proves a powerful
agent in the decomposition of mineral substances held
in solution, and of their consequent precipitation. An
earthen pitcher, containing several quarts of sulphate
of iron, had been suffered to remain undisturbed and
unexamined, in a corner of Mr Pepys' laboratory, for
about a twelvemonth. Some luckless mice had meanwhile fallen into it and been drowned; and when it
at length came to be examined, an oily scum and a
yellow sulphurous powder, mixed with hairs, were seen
floating on the top, and the bones of the mice discovered lying at the bottom; and it was found, that over
the decaying bodies the mineral components of the
fluid had been separated and precipitated in a dark-
coloured sediment, consisting of grains of pyrites and
of sulphur, of copperas in its green and crystalline form,
and of black oxide of iron. The animal and mineral
matters had mutually acted upon one another; and the
metallic sulphate, deprived of its oxygen in the process,
had thus cast down its ingredients. It would seem
that over the putrefying bodies of the fish of the Lower Old Red Sandstone the water had deposited, in like

Pepys: not the diarist, but Pepys 1811; Lyell: in *Elements of Geology*. Lyell 1838, pp. 85–86.

manner, the lime with which it was charged; and hence the calcareous nodules in which we find their remains inclosed. The form of the nodule almost invariably agrees with that of the ichthyolite within : it is a coffin in the ancient Egyptian style. Was the ichthyolite twisted half round in the contorted attitude of violent death ? the nodule has also its twist ; did it retain its natural posture ? the nodule presents the corresponding spindle form ; was it broken up, and the outline destroyed ? the nodule is flattened and shapeless. In almost every instance the form of the organism seems to have regulated that of the stone. We may trace, in many of these concretionary masses, the operations of three distinct principles, all of which must have been in activity at one and the same time. They are wrapped concentrically each round its organism ; they split readily in the line of the inclosing stratum, and are marked by its alternating rectilinear bars of lighter and darker colour ; and they are radiated from the centre to the circumference. Their concentric condition shows the chemical influences of the decaying animal matter ; their fissile character and parallel layers of colour indicate the general deposition which was taking place at the time ; and their radiated structure testifies to that law of crystalline attraction, through which, by a wonderful masonry, the invisible but well-cut atoms build up their cubes, their rhombs, their hexagons, and their pyramids, and are at once the architects and the materials of the structure which they rear.

Another and very different chemical effect of orga-

---

coffin ... style: a mummy case or sarcophagus.

nic matter may be remarked in the darker-coloured arenaceous deposits of the formation, and occasionally in the stratified clays and nodules of the ichthyolite bed. In a print-work the whole web is frequently thrown into the vat and dyed of one colour; but there afterwards comes a discharging process: some chemical mixture is dropped on the fabric; the dye disappears wherever the mixture touches; and in leaves, and sprigs, and patches, according to the printer's pattern, the cloth assumes its original white. Now the coloured deposits of the Old Red Sandstone have, in like manner, been subjected to a discharging process. The dye has disappeared in oblong or circular patches of various sizes, from the eighth of an inch to a foot in diameter; the original white has taken its place; and so thickly are these speckles grouped in some of the darker-tinted beds, that the surfaces, where washed by the sea, present the appearance of sheets of calico. The discharging agent was organic matter; the uncoloured patches are no mere surface films, for, when cut at right angles, their depth is found to correspond with their breadth, the circle is a sphere, the ellipsis forms the section of an egg-shaped body, and in the centre of each we generally find traces of the organism in whose decay it originated. I have repeatedly found single scales, in the ichthyolite beds, surrounded by uncoloured spheres about the size of musket-bullets. It is well for the young geologist carefully to mark such appearances,—to trace them through the various instances in which the organism may be recognised and identified, to those in which its last vestiges have disappeared. They are the hatchments of the geological world, and indicate that life once existed where all other record of it has perished.*

It is the part of the chemist to tell us by what peculiar action of the organic matter the dye was dis-

---

* Some of the clay slates of the primary formations abound in these circular uncoloured patches, bearing in their centres, like the patches of the Old Red Sandstone, half-obliterated nuclei of black. Were they too once fossiliferous, and do these blank erasures remain to testify to the fact ? I find the organic origin of the patches in the Old Red Sandstone remarked by Professor Fleming as early as the year 1830, and the remark reiterated by Dr Anderson of Newburgh in nearly the same words, but with no acknowledgment, ten years later. The following is the minute and singularly faithful description of the Professor :—

" On the surface of the strata in the lower beds, circular spots nearly a foot in diameter may be readily perceived by their pale-yellow colours, contrasted with the dark-red of the surrounding rock. These spots, however, are not, as may at first be supposed, mere superficial films, but derive their circular form from a coloured sphere to which they belong. This sphere is not to be distinguished from the rest of the bed by any difference in mechanical structure, but merely by the absence of much of that oxide of iron with which the other portion of the mass is charged. The circumference of this coloured sphere is usually well defined ; and at its centre may always be observed matter of a darker colour, in some cases disposed in concentric layers, in others of calcareous and crystalline matter, the remains probably of some vegetable or animal organism, the decomposition of which exercised a limited influence on the colouring matter of the surrounding rock. In some cases I have observed these spheres slightly compressed at opposite sides, in a direction parallel with the plane of stratification,—the result, without doubt, of the subsidence or contraction of the mass, after the central matter or nucleus had ceased to exercise its influence."—(Cheek's Edinburgh Journal, Feb. 1831, p. 82.)

---

hatchments: heraldic arms of a deceased person, placed in a special black frame as part of the mourning ritual; **Fleming**: Fleming 1831, p. 82; **Anderson**: presumably Anderson 1840, p. 192, or 1841, pp. 381–85; see Additional Note for p. 173.

charged in these spots and patches. But how was the dye itself procured ? From what source was the immense amount of iron derived, which gives to nearly five-sixths of the Old Red Sandstone the characteristic colour to which it owes its name ? An examination of its lowest member, the great conglomerate, suggests a solution of the query. I have adverted to the large proportion of red-coloured pebbles which this member contains, and, among the rest, to a red granitic gneiss, which must have been exposed over wide areas at the time of its deposition, and which, after the lapse of a period which extended from at least the times of the Lower Old Red to those of the Upper Oolite, was again thrust upwards to the surface, to form the rectilinear chain of precipitous eminences to which the hills of Cromarty and of Nigg belong. This rock is now almost the sole representative, in the north of Scotland, of the ancient rocks whence the materials of the Old Red Sandstone were derived. It abounds in hæmatic iron-ore, diffused as a component of the stone throughout the entire mass, and which also occurs in it in ponderous insulated blocks of great richness, and in thin thread-like veins. When ground down it forms a deep red pigment, undistinguishable in tint from the prevailing colour of the Sandstone, and which leaves a stain so difficult to be effaced, that shepherds employ it in some parts of the Highlands for marking their sheep. Every rawer fragment of the rock bears its hæmatic tinge; and were the whole ground by some mechanical process into sand, and again consolidated, the produce of the experiment would be undoubtedly a deep red sandstone. In an upper member of the lower formation,—that immediately over the ichthyolite beds,—different materials seem to have been employed. A white quartzy sand and a pale-coloured clay form the chief ingredients; and though the ochry-tinted colouring matter be also iron, it is iron existing in a different condition, and in a more diluted form. The oxide deposited by the chalybeate springs which pass through the lower members of the formation, would give to white sand a tinge exactly resembling the tint borne by this upper member.

The passage of metals from lower to higher formations, and from one combination to another, constitutes surely a highly interesting subject of inquiry. The transmission of iron in a chemical form, through chalybeate springs, from deposits in which it had been diffused in a form merely mechanical, is of itself curious; but how much more so its passage and subsequent accumulation, as in bog-iron and the iron of the Coal Measures, through the agency of vegetation? How strange, if the steel axe of the woodman should have once formed part of an ancient forest!—if, after first existing as a solid mass in a primary rock, it should next have come to be diffused as a red pigment in a transition conglomerate,—then as a brown oxide in a chalybeate spring,—then as a yellowish ochre in a secondary sandstone,—then as a component part in the stems and twigs of a thick forest of arboraceous plants,—then again as an iron carbonate slowly accumulating at the bottom of a morass of the Coal Measures,—then as a layer of indurated bands and nodules of brown ore underlying a

iron of the Coal Measures: blackband iron ore, sometimes found conveniently interspersed with the coal seams needed to calcine it, as in parts of the Midland Valley of Scotland.

seam of coal,—and then, finally, that it should have
been dug out, and smelted, and fashioned, and employed for the purpose of handicraft, and yet occupy, even at this stage, merely a middle place between the transmigrations which have passed, and the
changes which are yet to come. Crystals of galena
sometimes occur in the nodular limestones of the Old
Red Sandstone; but I am afraid the chemist would
find it difficult to fix their probable genealogy.

In at least one respect every geological history must
of necessity be unsatisfactory; and ere I pass to the
history of the two upper formations of the system, the
reader must permit me to remind him of it. There have
been individuals, it has been said, who, though they
could see clearly the forms of objects, wanted, through
some strange organic defect, the faculty of perceiving
their distinguishing colours, however well marked these
might be. The petals of the rose have appeared to them
of the same sombre hue with its stalk; and they have
regarded the ripe scarlet cherry as undistinguishable in
tint from the green leaves under which it hung. The
face of nature to such men must have for ever rested
under a cloud; and a cloud of similar character hangs
over the pictorial restorations of the geologist. The
history of this and the last chapter is a mere profile
drawn in black,—an outline without colour,—in short,
such a chronicle of past ages as might be reconstruct-
ed, in the lack of other and ampler materials, from
tombstones and charnel-houses. I have had to draw
the portrait from the skeleton. My specimens show
the general form of the creatures I attempt to describe, and not a few of their more marked peculiari-

ties ; but many of the nicer elegancies are wanting ;
and the " complexion to which they have come"
leaves no trace by which to discover the complexion
they originally bore. And yet colour is a mighty
matter to the ichthyologist. The " fins and shining
scales"—" the waved coats, dropt with gold"—the
rainbow-dyes of beauty of the watery tribes,—are connected often with more than mere external character.
It is a curious and interesting fact, that the hues of
splendour in which they are bedecked are, in some
instances, as intimately associated with their instincts
—with their feelings, if I may so speak—as the blush
which suffuses the human countenance is associated
with the sense of shame, or its tint of ashy paleness or
of sallow with emotions of rage, or feelings of a panic
terror. Pain and triumph have each their index of
colour among the mute inhabitants of our seas and
rivers. Poets themselves have bewailed the utter inadequacy of words to describe the varying tints and
shades of beauty with which the agonies of death
dye the scales of the dolphin, and how every various pang calls up a various suffusion of splendour.*
Even the common stickleback of our ponds and ditches
can put on its colours to picture its emotions.   There

* The description of Falconer must be familiar to every
reader, but I cannot resist quoting it. It shows how minutely
the sailor poet must have observed. Byron tells us how
                    " Parting day
      Dies like the dolphin, whom each pang embues
      With a new colour, as it gasps away,
      The last still loveliest, till—'tis gone, and all is gray."
Falconer, in anticipating, reversed the simile. The huge ani-

'complexion … come': adapted from 'Epitaph on James Quin, in Bath Cathedral'
(1785) by English actor and poet David Garrick (1717–79): 'To this complexion thou
must come at last' (i.e. all must die some day). The expression became proverbial; 'fins
… coats …': Milton, *Paradise Lost*, Book 7, lines 401 and 406, on God's creation of
fishes; **dolphin**: the fish *Coryphaena hippurus* (also known as dorado, mahi-mahi, and
dolphinfish); its colourful death-spasms were alluded to by many nineteenth-century
writers (Burnett 2013); **Falconer**: William Falconer (1732–69), Scottish poet and sea-
man; 'Parting day … gray': Byron, *Childe Harold's Pilgrimage*, Canto 4, stanza 29.

is, it seems, a mighty amount of ambition, and a vast
deal of fighting sheerly for conquest sake, among the
myriads of this pigmy little fish which inhabit our
smaller streams ; and no sooner does an individual
succeed in expelling his weaker companions from
some eighteen inches or two feet of territory, than
straightway the exultation of conquest converts the
faded and freckled olive of his back and sides into a
glow of crimson and bright green.   Nature furnishes
him with a regal robe for the occasion.   Immediately
on his deposition, however,—and events of this kind
are even more common under than out of the water,—
his gay colours disappear, and he sinks into his original and native ugliness.*

mal, stuck by the " unerring barb" of Rodmond, has been
drawn on board, and
  " On deck he struggles with convulsive pain.
   But while his heart the fatal javelin thrills,
   And flitting life escapes in sanguine rills,
   What radiant changes strike th' astonished sight !
   What glowing hues of mingled shade and light !
   Not equal beauties gild the lucid west
   With parting beams o'er all profusely drest.
   Not lovelier colours paint the vernal dawn,
   When orient dews impearl the enamell'd lawn,
   Than from his sides in bright suffusion flow,
   That now with gold empyreal seem to glow ;
   Now in pellucid sapphires meet the view,
   And emulate the soft celestial hue ;
   Now beam a flaming crimson on the eye,
   And now assume the purple's deeper dye.
   But here description clouds each shining ray,—
   What terms of art can Nature's powers display."

* " In the *Magazine of Natural History*," says Captain
Brown, in one of his notes to White's *Selborne*, " we have a

---

'On deck … display': from Falconer's epic poem *The Shipwreck* (1762), Canto 2;
Captain Brown … Selborne: the Scottish naturalist Thomas Brown (1785–1862),
not to be confused with the Brown quoted on p. 220, was one of several authors who
produced updated and annotated versions of the *Natural History* part of Gilbert White's
Natural History and Antiquities of Selborne of 1789 (see pp. A24–25, A33–34, A127 and A.
Secord 2013). Brown's edition was published in Edinburgh in 1833 (initially for 3s) and
was reprinted many times; the quoted passage is at Letter II of this edition (White 1836,
pp. 26–27 n.).

But of colour, as I have said, though thus important, the ichthyologist can learn almost nothing from

curious account of the pugnacious propensities of these little
animals. 'Having at various times,' says a correspondent,
'kept these little fish during the spring and part of the summer months, and paid close attention to their habits, I am
enabled from my own experience to vouch for the facts I
am about to relate. I have frequently kept them in a deal
tub, about three feet two inches wide, and about two feet
deep. When they are put in for some time, probably a day
or two, they swim about in a shoal, apparently exploring
their new habitation. Suddenly one will take possession of
the tub, or, as it will sometimes happen, the bottom, and will
instantly commence an attack upon his companions ; and if
any of them venture to oppose his sway, a regular and most
furious battle ensues. They swim round and round each other
with the greatest rapidity, biting (their mouths being well
furnished with teeth), and endeavouring to pierce each other
with their lateral spines, which on this occasion are projected.
I have witnessed a battle of this sort which lasted several
minutes before either would give way; and when one does
submit, imagination can hardly conceive the vindictive fury
of the conqueror, who, in the most persevering and unrelent-
ing way, chases his rival from one part of the tub to another,
until fairly exhausted with fatigue. From this period an interesting change takes place in the conqueror, who, from being a speckled and greenish-looking fish, assumes the most
beautiful colours ; the belly and lower jaws becoming a deep
crimson, and the back sometimes a cream-colour ; but generally a fine green ; and the whole appearance full of animation
and spirit. I have occasionally known three or four parts of
the tub taken possession of by these little tyrants, who guard
their territories with the strictest vigilance, and the slightest
invasion brings on invariably a battle. A strange alteration
immediately takes place in the defeated party ; his gallant
bearing forsakes him,—his gay colours fade away,—he becomes again speckled and ugly,—and he hides his disgrace
among his peaceable companions."

Geology. The perfect restorations of even a Cuvier are blank outlines. We just know by a wonderful accident that the Siberian elephant was red. A very few of the original tints still remain among the fossils of our north-country Lias. The ammonite, when struck fresh from the surrounding lime, reflects the prismatic colours, as of old ; a huge modiola still retains its tinge of tawny and yellow ; and the fossilised wood of the formation preserves a shade of the native tint, though darkened into brown. But there is considerably less of colour in the fossils of the Old Red Sandstone. I have caught, and barely caught, in some of the newly-disinterred specimens, the faint and evanescent reflection of a tinge of pearl ; and were I acquainted with my own collection only, imagination, borrowing from the prevailing colour, would be apt to people the ancient oceans in which its forms existed, with swarthy races exclusively. But a view of the Altyre fossils would correct the impression. They are inclosed, like those of Cromarty, in nodules of an argillaceous limestone. The colour, however, from the presence of iron and the absence of bitumen, is different. It presents a mixture of gray, of pink, and of brown ; and on this ground the fossil is spread out in strongly-contrasted masses of white and dark red, of blue and of purple. Where the exuviæ lie thickest, the white appears tinged with delicate blue,—the bone is but little changed. Where they are spread out more thinly, the iron has pervaded them, and the purple and deep red prevail. Thus the same ichthyolite presents, in some specimens, a body of white and

wonderful accident: i.e. the discovery, in 1799, of a frozen mammoth; see p. 156; **Altyre fossils**: i.e. kept at Altyre House in Lady Gordon Cumming's collection, rather than collected from the nearby Altyre Burn: Andrews 1982a, p. 72 n. 123, 1983. For the colourings described here, see the specimen in Figure 39 and the painting in Figure 61 (from an album by Gordon Cumming and her daughters).

plum-blue attached to fins of deep red, and with detached scales of red and of purple lying scattered around it. I need hardly add, however, that all this variety of colouring is, like the unvaried black of the Cromarty specimens, the result merely of a curious chemistry.

CHAPTER XIV.

THE curtain rises, and the scene is new. The myriads of the lower formation have disappeared, and we
are surrounded, on an upper platform, by the existences of a later creation. There is sea all around, as
before ; and we find beneath, a dark-coloured muddy
bottom, thickly covered by a dwarf vegetation. The
circumstances differ little from those in which the ichthyolite beds of the preceding period were deposited;
but forms of life, essentially different, career through
the green depths, or creep over the ooze. Shoals of
Cephalaspides, with their broad arrow-like heads and
their slender angular bodies, feathered with fins, sweep
past like clouds of cross-bow bolts in an ancient bat-

---

green depths: this and the description at the end of the chapter somewhat belie Miller's
earlier apology that his history is a mere 'outline without colour' (p. 250).

tle. We see the distant gleam of scales, but the forms
are indistinct and dim : we can merely ascertain that
the fins are elevated by spines of various shape and
pattern ; that of some the coats glitter with enamel ;
and that others,—the sharks of this ancient period,—
bristle over with minute thorny points. A huge crustacean, of uncouth proportions, stalks over the weedy
bottom, or burrows in the hollows of the banks.

Let us attempt bringing our knowledge of the present
to bear upon the past. The larger crustacea of the
British seas abound most on iron-bound coasts, where
they find sheltering places in the deeper fissures of
sea-cliffs covered up by kelp and tangle, or under the
lower edges of detached boulders, that rest unequally
on uneven platforms of rock, amid forests of the
rough-stemmed cuvy. We may traverse sandy or
muddy shores for miles together, without finding a
single crab, unless a belt of pebbles lines the upper
zone of beach, where the forked and serrated fuci
first appears, or a few weed-covered fragments of rock
here and there occur in groupes on the lower zones.
In this formation, however, the bottom must have
been formed of mingled sand and mud, and yet the
crustacea were abundant. How account for the fact ?
There is, in most instances, an interesting conformity
between the character of the ancient rocks, in which
we find groupes of peculiar fossils, and the habitats of
those existences of the present creation which these
fossils most resemble. The fisherman casts his nets
in a central hollow of the Moray Frith, about thirty
fathoms in depth, and draws them up foul with masses of a fetid mud, charged with multitudes of that

R

---

iron-bound: rocky.

curious purple-coloured zoophite the sea-pen, invariably an inhabitant of such recesses. The graptolite of the most ancient fossiliferous rocks, an existence of unequivocally the same type, occurs in
most abundance in a finely-levigated mudstone, for
it too was a dweller in the mud. In like manner,
we may find the ancient modiola of the Lias in habitats analogous to those of its modern representative the muscle, and the encrinite of the Mountain
Limestone fast rooted to its rocky platform, just as
we may see the Helianthoida and Ascidioida of our
seas fixed to their boulders and rocky skerries. But
is not analogy at fault in the present instance ? Quite
the reverse. Mark how thickly these carbonaceous
impressions cover the muddy-coloured and fissile sandstones of the formation, giving evidence of an abundant vegetation. We may learn from these obscure
markings, that the place in which they grew could
have been no unfit habitat for the crustaceous tribes.

There is a little land-locked bay on the southern
shore of the Frith of Cromarty, effectually screened
from the easterly winds by the promontory on which
the town is built, and but little affected by those of
any other quarter, from the proximity of the neighbouring shores. The bottom, at low ebb, presents a level plain of sand, so thickly covered by the green grass-
weed of our more sheltered sandy bays and estuaries,
that it presents almost the appearance of a meadow.
The roots penetrate the sand to the depth of nearly
a foot, binding it firmly together; and as they have
grown and decayed in it for centuries, it has acquired,
from the disseminated particles of vegetable matter, a

---

muscle: here, mussel; land-locked bay: Udale Bay west of Cromarty.

deep leaden tint, more nearly approaching to black than even the dark gray mudstones of Balruddery. Nor is this the only effect : the intertwisted fibres impart to it such coherence, that, where scooped out into pools, the edges stand up perpendicular from the water like banks of clay ; and where these are hollowed into cave-like recesses,—and there are few of them that are not so hollowed,—the recesses remain unbroken and unfilled for years. The weeds have imparted to the sand a character different from its own, and have rendered it a suitable habitat for numerous tribes, which, in other circumstances, would have found no shelter in it. Now, among these we find in abundance the larger crustaceans of our coasts. The brown edible crab harbours in the hollows beside the pools : occasionally we may find in them an overgrown lobster, studded with parasitical shells and zoophites,—proof that the creature, having attained its full size, has ceased to cast its plated covering. Crustaceans of the smaller varieties abound. Hermit-crabs traverse the pools, or creep among the weed ; the dark-green and the dingy hump-backed crabs occur nearly as frequently ; the radiata cover the banks by thousands. We find occasionally the remains of dead fish left by the retreating tide ; but the living are much more numerous than the dead ; for the sand-eel has suffered the water to retire, and yet remained behind in its burrow ; and the viviparous blenny and common gunnel still shelter beside their fuci-covered masses of rock. Imagine the bottom of this little bay covered up by thick beds of sand and gravel, and the whole consolidated into stone, and we have in it all the conditions of the de-

---

grass-weed: eelgrass *Zostera*.

posit of Balruddery,—a mud-coloured arenaceous deposit abounding in vegetable impressions, and inclosing numerous remains of crustaceans, fish, and radiata, as its characteristic organisms of the animal kingdom. There would be but one circumstance of difference: the little bay abounds in shells; whereas no shells have yet been found in the mudstones of Balruddery, or the gray sandstones of the same formation, which in Forfar, Fife, and Morayshires represent the Cornstone division of the system.

Ages and centuries passed, but who can sum up their number? In England the depth of this middle formation greatly exceeds that of any of the other two; in Scotland it is much less amply developed; but in either country it must represent periods of scarce conceivable extent. I have listened to the controversies of opposite schools of geologists, who from the earth's strata extract registers of the earth's age of an amount amazingly different. One class, regarding the geological field as if under the influence of those principles of perspective which give to the cottage in front more than the bulk and altitude of the mountain behind, would assign to the present scene of things its thousands of years, but to all the extinct periods united merely their few centuries; while with their opponents, the remoter periods stretch out far into the bygone eternity, and the present scene seems but a narrow strip running along the foreground. Both classes appeal to facts; and, leaving them to their disputes, I have gone out to examine and judge for myself. The better to compare the present with the past, I have regarded the existing scene merely as

---

radiata: Miller's identification of 'radiata' on p.150 was vague; **opposite schools**: Miller contrasts what are now termed young-earth and old-earth views regarding the age of the earth; he affects a neutral stance here, but has already shown that he assumes an immeasurably ancient earth (e.g. p.218, and see pp.A58–59); **extract registers**: alluding to 'The Garden' in *The Task* (1785) by William Cowper, meditating on pointless pursuits: 'Some drill and bore / The solid earth, and from the strata there / Extract a register, by which we learn, / That he who made it, and reveal'd its date / To Moses was mistaken in its age.'

a *formation*,—not as superficies, but as depth; and have sought to ascertain the extent to which, in different localities, and under different circumstances, it has overlaid the surface.

The slopes of an ancient forest incline towards a river that flows sluggishly onwards through a deep alluvial plain, once an extensive lake. A recent landslip has opened up one of the hanging thickets. Uprooted trees, mingled with bushes, lie at the foot of the slope, half-buried in broken masses of turf; and we see above, a section of the soil from the line of vegetation to the bare rock. There is an under belt of clay and an upper belt of gravel, neither of which contains any thing organic; and overtopping the whole we may see a dark-coloured bar of mould, barely a foot in thickness, studded with stumps and interlaced with roots. Mark that narrow bar: it is the geological representative of six thousand years. A stony bar of similar appearance runs through the strata of the Wealden : it too has its dingy colour, its stumps, and its interlacing roots ; but it forms only a very inconsiderable portion of one of the least considerable of all the formations, and yet who shall venture to say that it does not represent a period as extended as that represented by the dark bar in the ancient forest, seeing that there is not a circumstance of difference between them ?

We descend to the river side. The incessant action of the current has worn a deep channel through the leaden-coloured silt ; the banks stand up perpendicularly over the water, and downwards, for twenty feet together,—for such is the depth of the de-

---

The slopes of an ancient forest ... plain ... river ... morass: perhaps another composite scene. If an actual place (an 'appeal to facts'), it may be where the River Findhorn debouches onto the Moray coastal plain (see pp. 217–19); **clay and gravel**: 'diluvium'; **mould**: soil; **six thousand years**: many of a young-earth persuasion, following older chronological scholarship, believed that this was the full extent of the earth's history; Miller, like most geologists at the time, believed that it represented only the history of the human species (see p. A63); **stony bar**: almost certainly the Dirt-Bed in Dorset: Buckland 1836, vol. I, p. 489, vol. II, plate 57, fig. 1, and Lyell 1838, pp. 353–54.

posit. We may trace layer after layer of reeds, and flags, and fragments of driftwood, and find here and there a few fresh-water shells of the existing species. In this locality six thousand years are represented by twenty feet. The depth of the various fossiliferous formations united is at least fifteen hundred times as great.

We pursue our walk, and pass through a morass. Three tiers of forest-trees appear in the section laid open by the stream, the one above the other. Overlying these there is a congeries of the remains of aquatic plants, which must have grown and decayed on the spot for many ages after the soil had so changed that trees could be produced by it no longer; and over the whole there occur layers of mosses, that must have found root on the surface after the waters had been drained away by the deepening channel of the river. The six thousand years are here represented by that morass, its three succeeding forests, its beds of aquatic vegetation, its bands of moss, and the thin stratum of soil which overlies the whole. Well, but it forms, notwithstanding, only the mere beginning of a formation. Pile up twenty such morasses, the one over the other; separate them by a hundred such bands of alluvial silt as we have just examined a little higher up the stream; throw in some forty or fifty thick beds of sand to swell the amount; and the whole together will but barely equal the Coal Measures,—one of many formations.

But the marine deposits of the present creation have been perhaps accumulating more rapidly than those of our lakes, forests, or rivers? Yes, unquestionably, in

---

deposit. We may: should read 'deposit we may' (author's erratum, emended in 2nd edition); But the marine deposits …?: Miller imagines someone of a young-earth persuasion suggesting this to explain away the depth of the deposits.

friths and estuaries, in the neighbourhood of streams that drain vast tracts of country, and roll down the soil and clay swept by the winter rains from thousands of hill-sides ; but what is there to lead to the formation of sudden deposits in those profounder depths of the sea, in which the water retains its blue transparency all the year round, let the waves rise as they may ? And do we not know that along many of our shores the process of accumulation is well nigh as slow as on the land itself? The existing creation is represented in the little land-locked bay, where the crustacea harbour so thickly, by a deposit hardly three feet in thickness. In a more exposed locality, on the opposite side of the promontory, it finds its representative in a deposit of barely nine inches. It is surely the present scene of things that is in its infancy ! Into how slender a bulk have the organisms of six thousand years been compressed ! History tells us of populous nations, now extinct, that flourished for ages : do we not find their remains crowded into a few streets of sepulchres ? 'Tis but a thin layer of soil that covers the ancient plain of Marathon. I have stood on Bannockburn, and seen no trace of the battle. In what lower stratum shall we set ourselves to discover the skeletons of the wolves and bears that once infested our forests ? Where shall we find accumulations of the remains of the wild bisons and gigantic elks, their cotemporaries ? They must have existed for but comparatively a short period, or they would surely have left more marked traces behind them.

When we appeal to the historians, we hear much of a remote antiquity in the history of man : a more than

---

gigantic elks: *Megaloceros giganteus*, an extinct species of deer, until recently known as Irish Elk but now usually termed Giant Deer; see also p. 121.

twilight gloom pervades the earlier periods; and the
distances are exaggerated, as objects appear large in
a fog. We measure too by a minute scale. There
is a tacit reference to the three score and ten years of
human life; and its term of a day appears long to the
ephemera. We turn from the historians to the prophets, and find the dissimilarity of style indicating a
different speaker. Ezekiel's measuring-reed is graduated into cubits of the temple. The vast periods
of the short-lived historian dwindle down into weeks
and days. Seventy weeks indicated to Daniel, in the
first year of Darius, the time of the Messiah's coming.
Three years and a half limit the term of the Mahom-
medan delusion. Seventeen years have not yet gone
by since Adam first arose from the mould; nor has
the race, as such, attained to the maturity of even
early manhood. But while prophecy sums up merely
weeks and days when it refers to the past, it looks
forward into the future, and speaks of a thousand
years. Are scales of unequally-graduated parts ever
used in measuring different portions of the same map
or section,—scales so very unequally graduated, that
while the parts in some places expand to the natural
size, they are in others more than three hundred times
diminished? If not,—for what save inextricable confusion would result from their use,—how avoid the
conclusion, that the typical scale employed in the
same book by the same prophet represents similar
quantities, whether applied to times of outrage, delusion, and calamity, or set off against that long and
happy period in which the spirit of evil shall be bound
in chains and darkness, and the kingdom of Christ

---

term: lifespan; **ephemera**: insects such as mayflies, which often live only for days after
they emerge from the pupa; **measuring-reed**: in one of the Bible's most difficult
prophetic passages, from the book of the prophet Ezekiel (40:5), Ezekiel has a vision of
a future temple at Jerusalem, much larger than any such structure known, in which a
mysterious man measures it using a six-cubit-long measuring-rod ('reed' in the King
James text); **different speaker**: see Additional Notes; **Daniel**: one of the visions of the
biblical prophet Daniel, in which he is told (Daniel 9:24), 'Seventy weeks are determined upon thy people and upon thy holy city to finish the transgression, and to make

shall have come ? And if such be the case,—if the present scene of things be thus merely in its beginning, —should we at all wonder to find that the formation which represents it has laid down merely its few first strata ?

The curtain again rises. A last day had at length come to the period of the middle formation ; and in an ocean roughened by waves and agitated by currents, like the ocean which flowed over the conglomerate base of the system, we find new races of existences. We may mark the clumsy bulk of the *Holoptychius* conspicuous in the group ; the shark family have their representatives as before ; a new variety of the *Pterichthys* spreads out its spear-like wings at every alarm, like its predecessors of the lower formation ; shoals of fish of a type more common, but still unnamed and undescribed, sport amid the eddies ; and we may see attached to the rocks below, substances of uncouth form and doubtful structure, with which the oroctyologist has still to acquaint himself. The depositions of this upper ocean are of a mixed character : the beds are less uniform and continuous than at a greater depth. In some places they consist exclusively of sandstone, in others of conglomerate ; and yet the sandstone and conglomerate seem, from their frequent occurrence on the same platform, to have been formed simultaneously. The transporting and depositing agents must have become more partial in their action than during the earlier period. They had their foci of strength and their circumferences of comparative weakness ; and while the heavier pebbles which compose the conglomerate were in the course

---

an end of sins' before the coming of the Messiah (identified by the Church with Christ, using calculations based on the principle of one prophetic day to a calendar year); **Mahometan** (p. 264): some exegetes saw the rise and fall of Islam predicted in part of Daniel's prophecy; **long and happy period** (p. 264): the Millennium, the future thousand-year reign of Christ on earth after the chaining of Satan (Revelation 20).

shark family: in this context, acanthodians; **oroctyologist**: oryctologist.

of being deposited in the foci, the lighter sand which composes the sandstone was settling in those outer skirts by which the foci were surrounded. At this stage, too, there are unequivocal marks, in the northern localities, of extensive denudation. The older strata are cut away in some places to a considerable depth, and newer strata of the same formation deposited unconformably over them. There must have been partial upheavings and depressions, corresponding with the partial character of the depositions; and, as a necessary consequence, frequent shiftings of currents. The ocean, too, seems to have lessened its general depth, and the bottom to have lain more exposed to the influence of the waves. And hence one cause, added to the porous nature of the matrix and the diffused oxide, of the detached, and, if I may so express myself, church-yard character of its organisms.

Above the blended conglomerates and sandstones of this band a deposition of lime took place. Thermal springs, charged with calcareous matter slightly mixed with silex, seem to have abounded, during the period which it represents, over widely-extended areas; and hence probably its origin. An increase of heat from beneath, through some new activity imparted to the Plutonic agencies, would be of itself sufficient to account for the formation. I have resided in a district in which almost every spring was charged with calcareous earth; but in cisterns or draw-wells, or the utensils in which the housewife stored up for use the water which these supplied, no deposition took place. With boilers and tea-kettles, however, the case was different. The agency of heat was brought

---

unconformably: see p. 14. Miller might have had in mind the section cut by the River Findhorn from Sluie northwards through red sandstones, then the Cothall limestone and finally the upper yellow limestones. This sequence features elsewhere in the book (pp. 141, 217, 261); church-yard character: see p. 156.

to operate upon these ; and their sides and bottoms were covered, in consequence, with a thick crust of lime. Now, we have but to apply the simple principles on which such phenomena occur, to account for widely-spread precipitates of the same earth by either springs or seas, which at a lower temperature would have been active in the formation of mechanical deposits alone. The temperature sunk gradually to its former state ; the purely chemical deposit ceased ; the waters became populous as before with animals of the same character and appearance as those of the upper conglomerate ; and layer after layer of yellow sandstone, to the depth of several hundred feet, were formed as the period passed. With this upper deposit the system terminated.

Though fish still remained the lords of creation, and fish of apparently no superior order to those with which the vertebrata began at least three formations earlier, they had mightily advanced in one striking particular. If their organization was in no degree more perfect than at first, their bulk at least had become immensely more great. The period had gone by in which a mediocrity of dimension characterized the existences of the ancient oceans, and fish armed offensively and defensively with scales and teeth scarcely inferior in size to the scales and teeth of the gavial or the alligator, sprung into existence. It must have been a large jaw and a large head that contained, doubtless among many others, a tooth an inch in diameter at the base. I may remark, in the passing, that most of the teeth found in the several formations of the system are not instruments of mastication, but, like those in most of the existing fish, mere hooks for penetrating slippery substances, and thus holding them fast. The rude angler who first fashioned a crooked bone, or a bit of native silver or copper, into a hook, might have found his invention anticipated in the jaws of the first fish he drew ashore by its means ; and we find the hook-structure as complete in the earlier ichthyolites of the Old Red Sandstone, as in the fish that exist now. The evidence of the geologist is of necessity circumstantial evidence, and he need look for none other ; but it is interesting to observe how directly the separate facts bear, in many examples, on one and the same point. The hooked and slender teeth tell exactly the same story with the undigested scales in the fœcal remains alluded to in an early chapter.

In what could this increase in bulk have originated ? Is there a high but yet comparatively medium temperature in which animals attain their greatest size, and corresponding gradations of descent on both sides, whether we increase the heat until we reach the point at which life can no longer exist, or diminish it until we arrive at the same result from intensity of cold ? The line of existence bisects on both sides the line of extinction. May it not probably form a curve, descending equally from an elevated centre to the points of bisection on the level of death ? But whatever may have been the cause, the change furnishes another instance of analogy between the progress of individuals and of orders. The shark and the sword-fish begin to exist as little creatures of a span in length ; they expand into monsters whose bodies

foecal: fæcal; The line of existence: see Additional Note for p. 268; They expand into monsters: Miller is of course speaking metaphorically, but we may see here why Robert Chambers found it so easy to make Miller's progressive creationism serve his own hypothesis of species transmutation in *Vestiges of the Natural History of Creation* (1844).

equal in hugeness the trunks of ancient oaks ; and thus has it been with the order to which they belong. The teeth, spines, and palatial bones of the fish of the Upper Ludlow Rocks are of almost microscopic minuteness ; an invariable mediocrity of dimension characterizes the ichthyolites of the Lower Old Red Sandstone ; a marked increase in size takes place among the existences of the middle formation ; in the upper the bulky *Holoptychius* appears ; the close of the system ushers in the still bulkier *Megalichthys* ; and low in the Coal Measures we find the ponderous bones, buckler-like scales, and enormous teeth of another and immensely more gigantic *Holoptychius,*— a creature pronounced by Agassiz the largest of all osseous fish.* We begin with an age of dwarfs,—we end with an age of giants. The march of Nature is an onward and an ascending march ; the stages are slow, but the tread is stately ; and to Him who has commanded, and who overlooks it, a thousand years are as but a single day, and a single day as a thousand years.

We have entered the Coal Measures. For seven for-

---

* There have been fish-scales found in Burdiehouse, five inches in length by rather more that four in breadth, and of such strength that, in their original state, a musket-bullet would have rebounded flattened from their enamelled planes, as from the scales of the crocodile or the hide of the rhinoceros. Of the gigantic *Holoptychius* of this deposit we have still much to learn. The fragment of a jaw, in the possession of the Royal Society of Edinburgh, which belonged to an individual of the species, is $18\frac{1}{2}$ inches in length ; and it is furnished with teeth, one of which from base to point measures five inches, and another four and a half.

---

another ... *Holoptychius*: now identified as rhizodontid fishes; **a thousand ... years**: quoted (in reverse) from 2 Peter 3:8 to dispel suspicions that a geologically vast timescale is unbiblical. A literal reading of this passage would convert the Creation-date of 4004 BC in the margins of printed Bibles to a period more than two thousand million years ago on the divine timescale. See also Additional Note to p. 241.

mations together,—from the Lower Silurian to the Upper Old Red Sandstone,—our course has lain over oceans without a visible shore, though, like Columbus in his voyage of discovery, we have now and then found a little floating weed, to indicate the approaching coast. The water is fast shallowing. Yonder passes a broken branch, with the leaves still unwithered ; and there floats a tuft of fern. Land, from the mast-head ! land ! land !—a low shore thickly covered with vegetation. Huge trees of wonderful form stand out far into the water. There seems no intervening beach. A thick hedge of reeds, tall as the masts of pinnaces, runs along the deeper bays, like water-flags at the edge of a lake. A river of vast volume comes rolling from the interior, darkening the water for leagues with its slime and mud, and bearing with it to the open sea, reeds, and fern, and cones of the pine, and immense floats of leaves, and now and then some bulky tree, undermined and uprooted by the current. We near the coast, and now enter the opening of the stream. A scarce penetrable phalanx of reeds, that attain to the height and well nigh the bulk of forest-trees, is ranged on either hand. The bright and glossy stems seem rodded like Gothic columns ; the pointed leaves stand out green at every joint, tier above tier, each tier resembling a coronal wreath or an ancient crown, with the rays turned outwards ; and we see atop what may be either large spikes or catkins. What strange forms of vegetable life appear in the forest behind ! Can that be a club-moss that raises its slender height far more than fifty feet from the soil ? Or can these tall palm-like trees be actually ferns, and these spreading

----

Columbus: the Genovese explorer Christopher Columbus, whose sea voyage in 1492 made him the first European to visit South and Central America; approaching coast: i.e. the first fossil evidence of life on land is provided by the Coal Measures, which we have been approaching throughout the book as we move upwards through the Old Red Sandstone system (which Miller took to be marine deposits); water-flags: flag irises; bearing … sea: to be deposited on the sea bed; it was then thought that some coal seams were marine deposits: e.g. Lyell 1838, p. 441; Gothic columns: as in a cathedral (see p. 93 and n.); wreath … crown: e.g. a Greek or Roman victor's olive wreath.

branches mere fronds? And then these gigantic reeds!—are they not mere varieties of the common horse-tail of our bogs and morasses, magnified some sixty or a hundred times? Have we arrived at some such country as the continent visited by Gulliver, in which he found thickets of weeds and grass tall as woods of twenty years' growth, and lost himself amid a forest of corn fifty feet in height? The lesser vegetation of our own country, its reeds, mosses, and ferns, seem here as if viewed through a microscope: the dwarfs have sprung up into giants, and yet there appears to be no proportional increase in size among what are unequivocally its trees. Yonder is a group of what seem to be pines,—tall and bulky, 'tis true, but neither taller nor bulkier than the pines of Norway and America; and the club-moss behind shoots up its green hairy arms, loaded with what seem catkins above their topmost cones. But what monster of the vegetable world comes floating down the stream, —now circling round in the eddies, now dancing on the ripple, now shooting down the rapid? It resembles a gigantic star-fish, or an immense coach-wheel divested of the rim. There is a green dome-like mass in the centre, that corresponds to the nave of the wheel or the body of the star-fish; and the boughs shoot out horizontally on every side, like spokes from the nave, or rays from the central body. The diameter considerably exceeds forty feet; the branches, originally of a deep green, are assuming the golden tinge of decay; the cylindrical and hollow leaves stand out thick on every side, like prickles of the wild rose on the red, fleshy, lance-like shoots of

---

continent visited by Gulliver: the fictional land of Brobdingnag, inhabited by gigantic animals, plants and humans, in the satirical novel which became known as *Gulliver's Travels* (1726) by the Irish author Jonathan Swift (1667–1745); **monster**: *Stigmaria* was then interpreted as a 'huge succulent water plant', perhaps with a floating lifestyle: Buckland 1836, vol. I, pp. 476–78, vol. II, pp. 95–96 and plate 56; Lyell 1838, pp. 434–35.

a year's growth, that will be covered two seasons
hence with flowers and fruit.   That strangely-formed
organism presents no existing type among all the numerous families of the vegetable kingdom.   There
is an amazing luxuriance of growth all around us.
Scarce can the current make way through the thickets
of aquatic plants that rise thick from the muddy bottom ; and though the sunshine falls bright on the
upper boughs of the tangled forest beyond, not a ray
penetrates the more than twilight gloom that broods
over the marshy platform below.   The rank steam
of decaying vegetation forms a thick blue haze, that
partially obscures the underwood; deadly lakes of
carbonic acid gas have accumulated in the hollows ;
there is silence all around, uninterrupted save by the
sudden splash of some reptile fish that has risen to the
surface in pursuit of its prey, or when a sudden breeze
stirs the hot air, and shakes the fronds of the giant
ferns or the catkins of the reeds.   The wide continent before us is a continent devoid of animal life,
save that its pools and rivers abound in fish and mollusca, and that millions and tens of millions of the infusory tribes swarm in the bogs and marshes.   Here
and there, too, an insect of strange form flutters among
the leaves.   It is more than probable that no creature
furnished with lungs of the more perfect construction
could have breathed the atmosphere of this early period, and have lived.

Doubts have been entertained whether the limestone of Burdiehouse belongs to the Upper Old Red
Sandstone or to the Inferior Coal Measures. And the
fact may yet come to be quoted as a very direct proof

---

infusory tribes: infusoria, i.e. protozoans, rotifers, algae, plankton and similar small
aquatic organisms; **hot air**: geologists agreed that the climate of the coal forests was
warm and humid: e.g. Buckland 1836, vol. I, pp. 464–65, 481–82, Lyell 1838, pp. 438–39,
Murchison 1839, part I, pp. 149–52; **carbonic acid gas**: carbon dioxide. Adolphe Brongniart (1829) suggested that the coal forests grew at a time of high atmospheric levels of
this gas (Rudwick 2008, p. 171). Miller seems to have taken up this idea, discussed for
example by Hibbert (1835, pp. 261–64), in order to emphasise in providentialist vein
that the world was not yet ready for the introduction of the 'higher' animals, as they

of the ignorance which obtained regarding the fossils of the older formation, at a time when the organisms of the most of the other formations, both above and below it, had been carefully explored. The Limestone of Burdiehouse is unequivocally and most characteristically a Coal-Measure limestone. It abounds in vegetable remains of terrestrial or lacustrine growth, and these, too, the vegetables common to the Coal-Measures,—ferns, reeds, and club-mosses. One can scarce detach a fragment from the mass, that has not its leaflet or its seed-cone inclosed, and in a state of such perfect preservation, that there can be no possibility of mistaking its character. If in reality a marine deposit, it must have been formed in the immediate neighbourhood of a land covered with vegetation. The dove set loose by Noah bore not back with it a less equivocal sign that the waters had abated. Now, in the Upper Old Red Sandstone, none of these plants occur. The deposit is exclusively an ocean deposit, and the remains in Scotland, until we arrive at its inferior and middle formations, are exclusively animal remains. Its upper member, "the yellow sandstone," says Dr Anderson of Newburgh, " does not exhibit a single particle of carbonaceous matter,—no trace or film of a branch having been detected in it, though, if such in reality existed, there are not wanting opportunities of obtaining specimens in some one of the twenty or thirty quarries which have been opened in the county of Fife, in this deposit alone." No two bordering formations in the geological scale have their boundaries better defined

s

would not survive. Miller's **lakes** were high concentrations of this dense gas collecting in depressions in the ground.

dove: according to the biblical narrative of the Deluge or Flood (Genesis 8:8–9), when the rains had ceased after forty days and Noah's Ark was floating on the floodwaters, Noah let a dove fly free, and when it brought back an olive branch he knew that land was emerging once more; **Anderson**: from Anderson 1841, p.388.

by the character of their fossils than the Old Red
Sandstone and the Coal-Measures.

We pursue our history no farther. Its after course
is comparatively well known. The huge sauroid fish
was succeeded by the equally huge reptile,—the reptile by the bird,—the bird by the marsupial quadruped.
And at length, after races higher in the scale of instinct had taken precedence in succession, the one of
the other, the sagacious elephant appeared, as the lord
of that latest creation which immediately preceded
our own. How natural does the thought seem which
suggested itself to the profound mind of Cuvier, when
indulging in a similar review! Has the last scene in
the series arisen, or has Deity expended his infini-
tude of resource, and reached the ultimate stage of progression at which perfection can arrive? The philosopher hesitated, and then decided in the negative,
for he was too intimately acquainted with the works
of the Omnipotent Creator to think of limiting his
power; and he could, therefore, anticipate a coming
period, in which man would have to resign his post
of honour to some nobler and wiser creature,—the monarch of a better and happier world. How well it is
to be permitted to indulge in the expansion of Cuvier's thought, without sharing in the melancholy of
Cuvier's feeling,—to be enabled to look forward to the
coming of a new heaven and a new earth, not in terror, but in hope,—to be encouraged to believe in the
system of unending progression, but to entertain no
fear of the degradation or deposition of man! The
adorable Monarch of the future, with all its unsummed

Cuvier … happier world: see Additional Notes; **new heaven … earth**: alluding to St
John's prophecy of a final blessed state in the biblical book of Revelation (21:1);
adorable Monarch: Jesus Christ, whose incarnation as God become man (thus **flesh
of our flesh**) embodies the apex (**unsummed perfection**) of life's progress towards its
Creator as seen in the fossil record.

perfection, has already passed into the heavens, flesh of our flesh, and bone of our bone, and Enoch and Elias are there with him,—fit representatives of that dominant race, which no other race shall ever supplant or succeed, and to whose onward and upward march the deep echoes of eternity shall never cease to respond.

THE END.

JOHNSTONE AND FAIRLY, PRINTERS, EDINBURGH.

Enoch and Elias: Enoch, an antediluvian patriarch in the Old Testament, and Elijah, an Old Testament prophet, famous for having apparently passed directly from earth to heaven without dying in the process. Genesis 5:24 on Enoch: 'And Enoch walked with God: and he was not; for God took him'; 2 Kings 2:11 on Elijah: 'Elijah went up by a whirlwind into heaven'. Hence, in a common Protestant interpretation, they are **there with him** (Christ), whereas other humans will have to wait until they are resurrected from the dead at the end of history; **race**: the human species as a whole (see pp. 102, 215, 240, 275).

BRITISH ASSOCIATION FOR THE ADVANCEMENT OF SCIENCE.

Wednesday, September 23, 1840.

SECTION C.—GEOLOGY AND PHYSICAL GEOGRAPHY.

Mr Lyell in the Chair.

" Mr Murchison gave an account of the investigations and discoveries of Mr Hugh Miller of Cromarty (now Editor of the *Witness*), in the Old Red Sandstone. Various members of a great family of fishes, existing only in a deposit of the very highest antiquity, had been discovered by Mr Miller, Dr Fleming, Dr Malcolmson, and other gentlemen. M. Agassi had found these fishes to be characterized by the peculiarity of not having the vertebral column terminated at the centre of the tail, as in the existing species, but at its extremity. He spoke in the highest terms of Mr Miller's perseverance and ingenuity as a geologist. With no other advantage than a common education, by a careful use of his means, he had been able to give himself an excellent education, and to elevate himself to a position which any man in any sphere of life might well envy. Mr Murchison added, that he had seen some of Mr Miller's papers on Geology, written in a style so beautiful and poetical, as to throw plain geologists like himself into the shade—(Cheers). The fish discovered by Mr Miller, one or two fine specimens of which were on the table, was yet without a name; and perhaps M. Agassiz, who would now favour them with a description of the class to which it belonged, would assign it one.

" M. Agassiz stated, that since he first saw, five or six years ago, the fishes of the old deposits, they had increased to such an extent as to enable them to connect them with one large geological epoch. This had been still farther established by their having been found in the same formation by Mr Murchison in Russia, and Mr Miller in Ross-shire. These fishes were characterized in the most curious way he had ever seen. After briefly adverting to their peculiarities, as illustrated by

---

He spoke in the highest terms: i.e. Murchison.

the specimens on the table, M. Agassiz proposed to call Mr Miller's the *Pterichthys Milleri*. In the course of a subsequent conversation, the learned Professor added, that in lately examining the eggs of the salmon, he had observed that in the fœtal state of these fishes they have that unequally-divided condition of tail which characterizes so large a portion of the fishes in the older strata, and which becomes so rare in the fishes of the cretaceous and post-cretaceous formations.

" Dr Buckland said he had never been so much astonished in his life by the powers of any man as he had been by the geological descriptions of Mr Miller, which had been shown to him in the *Witness* newspaper by his friend Sir C. Menteath. That wonderful man described these objects with a felicity which made him ashamed of the comparative meagreness and poverty of his own descriptions in the *Bridgewater Treatise,* which had cost him hours and days of labour. He (Dr B.) would give his left hand to possess such powers of description as this man ; and if it pleased Providence to spare his useful life, he, if any one, would certainly render the science attractive and popular, and do equal service to Theology and Geology. It must be gratifying to Mr Miller to hear that his discovery had been assigned his own name by such an eminent authority as M. Agassiz ; and it added another proof of the value of the meeting of the Association, that it had contributed to bring such a man into notice.

" It was mentioned by Mr Murchison, that Lady Cumming Gordon has come forward as a distinguished and successful geological investigator in the north of Scotland."—*From the " Scottish Guardian."*

WORKS BY HUGH MILLER.

In small Octavo, price 7s. 6d.

SCENES and LEGENDS of the NORTH of SCOTLAND.

" A highly amusing and interesting book, written by a remarkable man, who will infallibly be well known." *Leigh Hunt's Journal.*

" A very pleasing and interesting book. The style has a

Sir C. Menteath: Charles Grenville Stuart Menteath or Menteith (1769–1847) of Closeburn in Dumfriesshire, advocate, landowner and collector of Carboniferous fossils from his Closeburn quarry: Menteath 1835; Knipe 1841. See Appendix 2; ***Bridgewater Treatise***: *Geology and Mineralogy* (Buckland 1836); ***Scottish Guardian***: see Appendix 2; **Cumming Gordon**: i.e. Gordon Cumming; see note for p.36.

3

purity and elegance which remind one of Irving, or of Irving's master, Goldsmith."—*Spectator.*

" Mr Miller is evidently a man of singular reflective powers, deep and enthusiastic feelings, and no small share of both humour and pathos."—*Chambers' Journal.*

ADAM & CHARLES BLACK, Edinburgh; LONGMAN & Co. London.

Fourth Edition, price 3d.

LETTER from ONE OF THE SCOTCH PEOPLE to the Right Hon. LORD BROUGHAM AND VAUX, on the Opinions expressed by his Lordship in the Auchterarder Case.

" It strikes a chord to which thousands of Scottish hearts will vibrate in unison."—*Presbyterian Review.*

" Over and above the judicial arguments in the Reports of the Auchterarder and Lethendy cases, the Church question has been discussed in a great variety of pamphlets, some of them very long and very able, others of them very long without being particularly able, and one of them particularly able without being long ; I mean the elegant and masculine production of Hugh Miller, entitled, ' A Letter to Lord Brougham.' " —*Church Principles Considered in their Results, by W. E. Gladstone, Esq. M.P.*

JOHN JOHNSTONE, Hunter Square, Edinburgh.

Second Edition, price 4d.

The WHIGGISM of the OLD SCHOOL, as exemplified in the PAST HISTORY and PRESENT POSITION of the CHURCH OF SCOTLAND.

" The first observation which strikes us with regard to this production is the admirable manner in which it is written. As a mere literary effort it must rank very high, for the purity and excellence of language, and the force and liveliness of its illustrations. Mr Miller is a master of English."—*Inverness Courier.*

JOHN JOHNSTONE, Hunter Square, Edinburgh.

DIRECTIONS TO THE BINDER.

ADDITIONAL NOTES

P. 11: thunder-bolts ... cattle: one of Miller's important records of local folklore, here confirming that 'thunderbolts' were not interpreted only as flint arrowheads.[1] Elsewhere Miller mentions that these local belemnites were administered powdered in water.[2] In Scottish and northern Irish folklore more generally, some ailments of cattle and humans were thought to result from the victim being 'elf-shot' with a fairy arrow, whether literally or metaphorically. The arrowheads themselves were widely thought in these areas to have healing properties, preserving their bearers from being elf-shot, or (as here) healing afflicted cattle. Their curative power makes sense on the principle of sympathetic magic, the 'hair of the dog that bit you'.[3] The notion of cattle being wounded by the fairies was also associated in parts of Lincolnshire with the belemnites found there.[4]

P. 18: disputed: during the 1830s the Old Red Sandstone had consensually been treated as part of the Carboniferous, but by the end of the decade it was raised to the same rank of system.[5] This took place against the background of a major controversy centring on the rocks of Devon and Cornwall, in which Murchison and Sedgwick opposed Henry De la Beche, the Director of the Geological Survey. Eventually Sedgwick (see p. 79) and Murchison (see Additional Note for p. 19) resolved it by their recognition in south Devon of rocks of the same age but very different type from the Old Red Sandstone, those Devon rocks being limestones rich in fossil shells and corals. Accordingly, in April 1839, they defined the Devonian System of which the Old Red Sandstone was the local expression in Scotland and the Welsh Marches.[6]

---

1   Taylor and Cheape 1999; Goodrum 2008, pp. 491–93.
2   Miller 2022 [1858], pp. 374–75.
3   Henderson and Cowan 2001, pp. 77–79 and 93–94; the fundamental study is now Hall 2005. For other Scottish examples of this belief, see Sibbald 1683–84, part 2, book 4, section 2 (*De fossilibus et murinis*), chapter 7, s.v. *lapis lanceæ cuspidem referens*; Brand 1813, vol. II, pp. 336–39; Dalyell 1834, pp. 157, 351–58 and 539; Campbell 2005, pp. 14 and 223.
4   Simon Knell, pers. comm. 2002.
5   Buckland 1836, vol. II, plate 66; Lyell 1838, pp. 421, 452; Murchison 1839, part I, pp. 169–70; Rudwick 1985, pp. 83–84.
6   Sedgwick and Murchison 1839; Rudwick 1985, pp. 280–87; 2008, pp. 451–56.

P. 19: **celebrated author**: Roderick Murchison (1792–1871), ex-army officer, and one of the most prominent members of the Geological Society of London, which then dominated British geology; at this time he was a gentleman-geologist of independent means, but in later years he became Director-General of the Geological Survey. He was primarily a stratigrapher rather than palaeontologist but was a major exponent of the use of fossils, as well as field relationships, to correlate strata, and so relied heavily on palaeontological collectors and taxonomists such as Miller and Agassiz (see pp. A46–47). His particular focus was on working out the sequence of the older rocks, to which his book *The Silurian System* (1839) was a major contribution. The Old Red Sandstone fishes were therefore of crucial importance to his work.[7]

P. 19: **Lyell … elementary work**: Charles Lyell (1797–75), a Scot brought up in England, geologist, leading member of the Geological Society of London and author of the seminal *Principles of Geology* (1830–33, with cheaper editions during the 1830s), an important intervention in scientific and public debates on how the science should be practised and why it was important.[8] His *Elements of Geology* (1838 and later editions) outlined the geological succession of strata, recasting material from the last volume of *Principles* in a more elementary manner.[9] Miller quoted Lyell (1838, pp. 452–54) with deletion of references and illustrations, but put 'northern' rather than 'southern half of Forfarshire', and 'northern' for 'southern borders of the Grampians'. Lyell was writing before Agassiz's naming of the fish as *Holoptychius*.[10] On the implications of this reference for Miller's target audience, see pp. A111–12.

P. 46: **introducing to … geologists**: *Pterichthys* fossils were first shown to other geologists either during John Malcolmson's and John Fleming's visits to Cromarty in late 1837 and (probably) end 1837 to early–mid 1838 respectively (see pp. 128, 143), or when Miller sent details and specimens to Paris about June 1838. Even then *Pterichthys* still seems to have been confused with *Coccosteus*.[11] This Paris letter (referred to on p. 130 as including a 'rude drawing', but not mentioning specimens) must have been sent sometime around March to May 1838, given Malcolmson's travel there in April, and the consignment of specimens which Miller sent at the end of May or start of June.[12] The letter might have been sent with this consignment, or an earlier communication, and this range of dates allows for that and for Miller writing early to Paris in anticipation of Malcolmson's arrival. Either interpretation

7   Andrews 1982a; Rudwick 1985; Secord 1986; Bonney and Stafford 2009.
8   Rudwick 2008, pp. 253–390; 2012; Secord 2014, pp. 138–72.
9   Rudwick 1990, p. liv.
10 Agassiz in Murchison 1839, part II, pp. 599–601, plate 2[bis], fig. 1; Andrews 1982a, pp. 18–19, 62.
11 Andrews 1982a, pp. 22–23.
12 See Andrews 1982a, pp. 22–23.

would indeed give about seven years, since the first *Pterichthys* was found in August 1830 (pp. 110, 117); on p. 117 Miller dated its true discovery to his realisation that it was something distinct, i.e. 'more than a twelvemonth later', but the wording 'first laid open to the light' can only mean field collecting.

P. 173: Anderson: Miller was evidently annoyed with Anderson for having misrepresented the views of Miller's friend and colleague Fleming, casting a slur on Fleming's competence.[13] He exposed this crime in his enormous footnote on pp. 166–170, which persisted into all later editions, even after resetting of the type. It is hard to make full sense of the degree of annoyance displayed by Miller here without knowing for certain which piece caused the offence (p. 167).[14] Of the pieces we do know about, two are mentioned in Miller's *Witness* articles which treat Anderson much more courteously than in the book. Miller used a discussion of oxidation-reduction chemistry in sandstone (p. 247 n.) to take another swipe at Anderson, but again without specifying which publication was causing offence; in this case, we have a likely candidate, in which Anderson seems guilty at most of failing to acknowledge Fleming's prior work.[15] However, this was bad practice and all the more potentially sensitive as Anderson had previously failed to acknowledge use of Fleming's work in published notices, and Fleming had published more correct interpretations of the fossils in question, albeit without drawing explicit attention to Anderson's blunders.[16]

It was probably Miller's attitude that changed rather than anything which Anderson published. He may have learned more about the background of Anderson's history with Fleming, or simply did not have room in his *Witness* articles to indulge himself. But another explanation for Miller's new aggressiveness is the ecclesiastical polemics over religious freedom which led to the Disruption.[17] Anderson was a Moderate and on the opposing side, and Miller did not always keep his church politics separate from his science or his literature,[18] nor was he known for pulling his punches in ecclesiastical or religious controversies. At about the time Miller wrote the *Witness* articles mentioned above, he was attacking Anderson for criticising a

13 See p. 166; pp. A53 and A134–35 n. 98; and Andrews 1982a, pp. 29–31.
14 Anderson 1837, 1840 (but apparently issued as a separate partwork in 1837: Anon. 1837b), 1841; [Miller] 1840h; Andrews 1982a, pp. 18–19, 25–31. Andrews (1982a, p. 62 n. 73) points out that the nominally 1841 essay was available in 1840, as Miller's *Witness* article confirms. At the time of writing, Fife newspapers of this period were poorly covered by online databases.
15 The more likely candidate is Anderson 1841, pp. 381 and 385, rather than Anderson 1840, p. 192.
16 Andrews 1982a, p. 12.
17 As suggested, tentatively, by Andrews 1982a, pp. 29–31, and rather more firmly by Waterston 2002b, pp. 88–89.
18 As in his controversy with the journalist Thomas Aird: Anon. 1848, pp. 110–11.

candidate minister in a nearby parish, partly because of this candidate's opposition to patronage.[19] And later, when Miller reported the moment which marked the Disruption itself – the Evangelicals' walkout from the Church of Scotland's General Assembly on 18 May 1843 to form the Free Church of Scotland – he would again set his sights on Anderson, calling him '*Moderate* science personified', repeating the beetle anecdote and dismissing him as 'a dabbler in geology', as opposed to the true science ('science, not *moderate*') of a good Evangelical such as David Brewster.[20] However, Kirk politics cannot be the full explanation, as geologically minded ministers of the Established Kirk did not automatically draw Miller's fire.[21]

Miller's attacks may seem unreasonable or ungentlemanly today, but this is from the viewpoint of an era accustomed to the convention of methodological naturalism in scientific debate (on which see pp. A55–56). In fact, these attacks raise interesting questions about the interplay of science and religion in his time. Miller had strong views on the vital need to ensure the best ministers for parishes.[22] The blurring of the boundary between scientific and ecclesiastical credibility as Miller perceived them, at least in the case of Anderson,[23] suggests that Miller genuinely saw the Fleming affair as reflecting badly on Anderson's personal morality and fitness to be a minister. Moreover, like many contemporaries, Miller would not have drawn a sharp boundary between religion and science, especially geology with its strong natural-theological content. Misconduct in science could be equated with misconduct in religion. Even if we set religious and ecclesiastical concerns to one side, natural-history authors in the early nineteenth century saw it as their duty to call out incompetent observers.[24]

Fleming's own silence is curious, for he had been notably combative in debate with other geologists back in the 1820s and early 1830s, partly in defence of what he perceived as his own rights of priority.[25] Adam Sedgwick once referred to Fleming

---

19 The local newspaper attacked the *Witness* in turn: Anon. 1840h.

20 [Miller] 1843, quoted and discussed in Brooke 1999, p. 30, and Secord 2000, p. 287.

21 For instance, he wrote positively in the mid-1840s about Charles Clouston of Sandwick in Orkney: see Miller 2022 [1858], p. 464.

22 Miller 1870, pp. 232–39; Henry 1996, p. 173; Macleod 1996, pp. 201–3.

23 As pointed out by Brooke 1999, p. 30.

24 Compare how the botanists Dawson Turner and L. W. Dillwyn flagged up serious errors in another botanist's work in 1805 (Secord 2011, pp. 289–92).

25 Bryson 1861; Burns 2007; Moore 2009. Burns (2007, p. 223) noted the collapse in Fleming's published output in the 1830s and suggested that Fleming had somewhat lost interest in science in the wake of his son's death in 1832 and his own move to a more demanding parish. But Fleming remained scientifically active, securing an appointment as professor of natural history at King's College, Aberdeen in 1834, and later taking up a similar post at the Free Church of Scotland College at Edinburgh in 1845. Perhaps the demands of his parish and university classes in the 1830s, as well as the bereavement, contributed to a slowing of Fleming's rate of publication.

metaphorically tomahawking and scalping William Buckland during the diluvial debates of the 1820s, in much the same terms as Free Church ministers who described Miller on the warpath against rival newspaper editors.[26]

Anderson later attempted to set the record straight, with ponderous sarcasm, in his book *The Course of Creation* (1850), which nevertheless also contains a generous endorsement of Miller's *Foot-Prints of the Creator*.[27]

P. 213: Donald Calder and the fairies: the story of Donald Calder and the fairies is a variant of the tale best known as 'Rip van Winkle' (after a literary version by Washington Irving published in 1819). This tale was widespread in Gaelic Scotland, and the drinking of whisky plays an important role in several variants.[28] Over-indulgence was certainly standard procedure after a market day in this period, at least for the menfolk. In this case, however, David Alston has suggested that Calder made up the story to explain his delayed return from Fortrose market as a result of dalliance with a local woman.[29] But, as always when reading Miller, one must consider the possibility of deadpan irony on his part, and for that matter on the part of the tale's original tellers, who were probably his then long-dead grandparents or great-aunts. So one wonders if it was invented ironically, or perhaps euphemistically, by Calder's fellow Cromarty denizens. The joke would be all the better if the story was, on the surface, innocuous enough to be told to youngsters such as Miller.

To be sure, Miller had further political and personal motivations for telling this story. Donald Calder's son James was a Radical in politics, a noted member of the Convention of Delegates of Friends of the People, and a witness at the 1793 trial of Maurice Margarot, one of the group known as the 'Scottish Political Martyrs'.[30] Miller's *Scenes and Legends* had included a sardonic but not unsympathetic portrayal of James as the enthusiastic local democrat, which was criticised as 'improper levity' in the Radical *Tait's Magazine*.[31] Almost certainly as a result of the *Tait's*

---

26 As quoted by Bryson 1861, p. 670, and Taylor 2022a, pp. 75–76.

27 Anderson 1850, pp. 61–67, and 335 n., discussed in Andrews 1982a, 30–31, and Secord 2000, p. 287.

28 Campbell 2005, pp. 33–36.

29 Alston 1996, p. 223.

30 Anon. 1837a, p. 9, 1847a.

31 Miller 1835, pp. 395–403; Anon. 1837a, p. 9, which contained the criticism; 1847a, p. 270. The po-faced tone of the *Tait's* criticism was perhaps understandable in the context of the government's harsh treatment of Radicals; even at Cromarty, as Miller notes, James was persecuted by a false allegation of importing weaponry with a view to insurrection, which fortunately was easily and publicly refuted. But we have been unable to substantiate the magazine's later claim in 1847 that *Scenes and Legends* called Calder a 'wild and furious democrat – the terror of the surrounding country'. The expression is given in quotation marks, implying that those were Miller's actual words, but the phrasing and tone adopted in *Scenes and Legends* itself (in 1835) are quite different.

report, James Calder threatened a libel suit over the description of him as a 'democrat', much to the surprise of Miller and his publisher Adam Black.[32] Black wrote supportively to Miller, who privately replied, dishing the hitherto hidden dirt. James's father Donald had been a prominent if somewhat Pharisaical lay member of the Cromarty kirk congregation, active there and in the neighbouring parish. But he lost his reputation when his visits to the other parish turned out to be to a woman whom he got pregnant, and he ended up a 'foul mouthed libidinous old man'.[33] This may explain the (doubtless ironic) dissonance in Miller's account evident to any local: that a man purportedly so active in the work of the Kirk should publicly admit to hearing the Good People. Miller told Black that James had left Cromarty about eight to ten years before Miller was born, i.e. around 1792–94, not long before Donald's own departure as a result of the public revelation of his liaison.[34] David Alston notes that the Cromarty Kirk Session records indeed show a paternity suit brought against Donald in 1796 by the widow Betty Stuart who claimed her first 'connection' with Calder at the Tain market in 1793.[35]

And there might be still more to the Calder story. Rather curiously, an 1847 account of James described Donald as 'deeply tinged with religious enthusiasm' and having had a prophetic dream as a child which later led him to start his business across the Firth of Cromarty, thus establishing the town of Invergordon.[36] Doubtless this was a response to the perceived rise of Invergordon at Cromarty's expense. But was it his own self-aggrandising claim, or a piece of embryo folklore told by others? At any rate, a recent history of the Firth firmly credits Invergordon's development to the landowners, and does not mention Calder.[37]

Be that as it may, most readers of *The Old Red Sandstone* (just like most children of Calder's time) would probably have taken the story of Calder and the fairies as just another variant on a well-known traditional pattern, and in this respect it is another illustration of the survival of fairy beliefs in eighteenth-century Scotland. This tale, like the other fairytales which accompany it here, should remind us, not of the arguably mythical Miller haunted by superstition throughout his life, but of

32   Miller-Black correspondence and associated documents, 23 January 1837 onwards, HMLB items 176–79.
33   Miller to Black, 16 February 1837, HMLB item 179. This letter also confirms that Donald Calder was present in *Scenes and Legends* as a 'Jacobin merchant' and a shopkeeper who refused to change a pound note, apparently out of fear of paper money losing its value in the coming revolution (Miller 1835, pp. 395, 402).
34   Anon. 1847a gives *c.*1792 for James's departure.
35   Alston 1996, p. 223, pers. comm. 2011. Kirk Session (Scots): committee of elders running a Presbyterian parish.
36   Anon. 1847a, p. 269.
37   Ash 1991.
38   Alston 1996; Cowan 2003, pp. 78–79; Henderson 2003, pp. 94–96 (who gives more weight to Miller's putative belief in fairies); Taylor 2022a, pp. 48–49, 142–45. See also pp. A78–81.

his indulgent and even-handed yet subtly ironic treatment of local lore, and his importance as a tradition-bearer in his own right.[38]

P. 217: wooded eminence: this viewpoint is difficult to reconstruct today owing to changes in the heron population, forestry and pasturage, exacerbated by Miller's characteristically fluid approach to scenic word-painting. The description probably conflates several real-life viewpoints along the east bank of the River Findhorn between the Long Rack (as marked on the current Ordnance Survey 1:25 000 map just 500m north-east of the farm of Mains of Sluie) and the cliff above Scur Pool a mile's walk to the north. Neither of the two named heronry sites on this stretch marked on the modern map (Heronry Pool and Darnaway Heronry) quite fits the topography described here. This description seems to be based on a much earlier trip that Miller had made before fossils were located in this area. Miller wrote to Patrick Duff on 27 January 1839 that 'I have seen the section exhibited by the Findhorn and the limestones of Cot-hall', but did not mention finding any fossils.[39] He was probably referring here to one or more of his visits to his literary friend Helen Dunbar of Boath in Forres in 1833 and 1834. Miller's letter to her of 24 July 1833 contains some similar elements, such as the herons, but these are much more briefly treated than in the book. The letter also reveals that, by July 1833, he already had in his mind a vivid series of scenic recollections from walking this stretch of the Findhorn 'rolled up in a recess of my mind, like a web of tapestry, and when I but will it, it unfolds scene after scene'. The description in the book may well represent more than one such scene, conflated in imagination (either in memory or by literary design) six or seven years later.[40]

P. 240: change of species: this passage is potentially confusing in that it deals with two levels of catastrophe, a loaded term in the geology of the time. Miller starts by discussing the fossil fish beds at Cromarty which he has already explained as evidence of a purely local disaster (compare pp. 232–36), although it might have led to local replacement of fish species by different ones (much as a forest fire leads to the replacement of mature trees, initially by rosebay willowherb and other weeds). However, after discussing a series of such local episodes, he then moves on to

---

39 ELGNM Geology Letter G3/4. On the Cothall 'limestones', see note to p. 218.
40 The letter of 24 July 1833 is HMLB item 64, printed with minor divergences in Bayne 1871, vol. I, pp. 334–43 (for the herons and the tapestry analogy see p. 340). For evidence that Miller visited Forres in 1834, see ibid., pp. 378–80, but Miller does not mention having visited the Findhorn on that occasion; see also Miller 1854a, p. 436, 1993, p. 434 for a brief reference to these visits, and see also Appendix 2. He returned to the Darnaway-Sluie stretch of the Find-horn in August 1844, as described in Chapter 12 of *The Cruise of the Betsey* (Miller 2022 [1858], pp. 196–98). On Miller's tendency to combine multiple viewpoints or visits into single recollected scenes, see Taylor 2022b, p. A46.

discussing the extinction of species rather than populations. It is possible that he was thinking of a global catastrophe, but this is not at all clear, and Miller here emphatically refuses to speculate about the processes involved. See the discussion in pp. A61–63.

P. 241: 'all things have continued …': adapted from a much-quoted biblical passage in 2 Peter 3: 4, which foretells that during the last days of the earth, 'scoffers' will claim that Christ will never return in the prophesied apocalypse, because things have always been the same and always will be ('since the fathers fell asleep, all things continue as they were from the beginning of the creation'). Peter takes such people to task for ignoring God's catastrophic interventions of the past (the Deluge) and future (the burning of the world in the Last Judgement). It is in this context that Peter reminds his audience that God operates on a vast timescale: 'one day is with the Lord as a thousand years, and a thousand years as one day' (2 Peter 3: 8), a phrase quoted on p. 269 in allusion to the length of the geological timescale. Miller echoes Peter's mockery of sceptics who doubt the possibility of a blessed afterlife on the basis of a few thousand years of unhappy human history. He then extends the analogy by suggesting that, had an intelligent being witnessed life on this planet up to and including the age of fishes, the vastly extended timescale involved would have made it natural (and more excusable) for that being to infer that no higher life-forms would ever be created. But geology confirms that God can still pull a surprise after billions of years of apparent stasis, creating a new and higher order of beings to usher in a new geological epoch. The philosophical implications of Peter's argument for the concept of natural law had already been much debated by early modern philosophers, and Miller would later enter this debate more fully in *The Testimony of the Rocks*.[41]

P. 241: wonderful analogies: in the early nineteenth century, European anatomists and embryologists in the Romantic idealist tradition produced differing explanations as to how and why the individual embryo of a mammal appeared to pass sequentially through forms akin to the various 'lower' animal forms as it developed from the fertilised egg (i.e. during its ontogeny), whereas the embryos of 'lower' animals stopped developing at earlier points in the sequence depending on their place in the scale (i.e. on what an evolutionist would call their phylogeny). For German *Naturphilosophen* and French transcendental anatomists such as Étienne Serres, the developing embryo 'recapitulated' the adult forms of the lower animals in a progressive, linear sequence, whereas the German-Estonian embryologist Karl Ernst von Baer favoured a non-linear model of increasing specialisation in different directions depending on the animal and its functional needs. Von Baer argued against recapitulation by arguing that all animal embryos look alike in their young,

---

41   Miller 1857b, pp. 192–205.

undifferentiated forms, but that 'higher' animal embryos become variously differen-
tiated as they mature. Because of the emerging evidence for progression in the fossil
record, both views proved highly amenable to both creationist and evolutionist
explanations of the history of life and were much debated in Edinburgh, as elsewhere.[42] Recapitulationism was adopted by both the transmutationist Geoffroy
Saint-Hilaire, Serres's teacher, and the creationist Agassiz, while von Baer's approach
formed the basis of Richard Owen's natural philosophy both before and after his
adoption of evolutionism, as well as the later theories of Darwin.[43] Drawing on his
detailed fossil-fish research, Agassiz was especially influential in documenting the
threefold link between embryonic development, animals' places in the scale of
existence, and the history of life on earth, inevitably progressing towards its goal, the
creation of humans.[44] Miller's treatment of this question – including his emphasis
on God as divine artist – resembles Agassiz's views as published after *The Old Red
Sandstone*.[45] It might, however, have derived in part from hearing (and conversing
with) Agassiz just after the British Association meeting in 1840 (see p. 243), and also
from press coverage of Agassiz's paper entitled 'On the Development of the Fish in
the Egg' with the aim of elucidating 'any relation between the forms of fish of the
present day, during the successive stages of their development, and the permanent
[presumably meaning 'adult'] forms of fish found in the older strata of the earth'.
This was not published until 1841, and then only as a brief and unrevealing abstract.[46] However, press coverage confirms its potential relevance. As quoted in the
publisher's end matter of Miller's book (end matter p. 2), the *Scottish Guardian*
noted that Agassiz had observed that foetal salmon in the egg 'have that unequally
divided condition of tail which characterises so large a portion of the fishes in the
older strata, and which becomes so rare' in later fishes.[47] In other words, the embryonic salmon retained the dorsoventrally asymmetrical tail of geologically earlier
fishes, as in Miller's finds. Miller's views would, in any case, be developed more fully
(and draw on Owen as well as Agassiz) in chapter 5 of *The Testimony of the Rocks*.[48]

---

42 Gould 1977; Rehbock 1983; Desmond 1989, pp. 52–59; Bowler 2009, pp.120–29. It is not clear
    how far Miller was involved in the Edinburgh scientific community, at least in these early
    days of the *Witness*; certainly he later came to be an active member of the Royal Physical Soci-
    ety of Edinburgh, which was the city's main general scientific society, outside the more exclu-
    sive Royal Society of Edinburgh. See Taylor 2022a, pp. 97–98.
43 Desmond 1989, pp. 337–72; Rupke 2009, pp. 83–134. As Rupke documents, the key shift in
    Owen's attitude towards evolution came at some time in the early to mid-1840s.
44 Gould 1977, pp. 63–68.
45 See Bowler 2009, pp.122–23.
46 Agassiz 1841.
47 Anon. 1840i, p. [2]. Agassiz's salmon work also featured in the discussion of a paper by Martin
    Barry on mammalian embryology (Anon. 1840r, pp. 847, 849).
48 Miller 1857b. For comments on Miller's place in the idealist tradition of fossil anatomy, see
    Livingstone 1984, pp.11–13.

P. 243: Butler ... Author of Revelation: Miller alludes to a classic work of Christian apologetics, *The Analogy of Religion, Natural and Revealed, to the Constitution and Course of Nature* (1736) by the English theologian Joseph Butler (1692–1752), whose work remained very influential in the period 1830–60.[49] In this work Butler had compared the character of divine government as deduced from the natural world with that revealed in the Bible ('Revelation'), concluding (against the deists) that the same God is visible in both; Miller is suggesting that geological evidence reinforces this view, a position he would develop more fully in *The Testimony of the Rocks*.

P. 264: different speaker: i.e. God or his angels. Miller's argument in the rest of this paragraph may be paraphrased as follows. Humans think 70 years is a long time, but the voice of God as mediated through the biblical prophets speaks of centuries of human history in the language of 'days' and 'weeks' (one prophetic 'day' being interpreted as equivalent to one calendar year in most Protestant exegesis at this time). Yet some biblical prophecies speak of the reign of Christ at the end of human history as lasting a thousand years. If we assume that the same day-year principle was used here, too, then these future periods dwarf the span of human history. We should thus not be surprised to find our history dwarfed by pre-human time as well. With this sleight of hand, Miller uses a literalistic and mainstream reading of biblical prophecy to undermine the plausibility of a young-earth geohistory, implying that the popular 'literalist' view in fact ran *counter* to a careful literal understanding of Scripture.

P. 268: line ... of existence: this is part of a wider suggestion that, even if a new group of animals came into existence at a small size, as usually happened, it would tend to form larger and larger members till it reached the optimum set by the ambient temperature at the time. Miller is evidently thinking about the way in which climate, through environmental temperature (whether average or extreme), might control the body size of animals and plants. He is suggesting that there is a certain optimal temperature for each organism when its size is greatest, with the size falling off on either side (the line of existence); at some point the climate is too hot or too cold for the organism to survive at a viable body size (cut off by the **level of death**). On the face of it Miller seems to be visualising a graph, the line of existence forming what we would today call a normal curve on a plot of body size (vertical axis) against temperature (horizontal axis). But it is not immediately clear how to draw a meaningful 'line of death', unless this is simply a more or less horizontal line corresponding to the minimum viable size, which would indeed **bisect** the line of existence on both sides. Also, such a notion seems more convincing for plants

---

49 Yule 1976, pp. 165–66.

(which can grow in stunted form) than animals. But, as with the brigantine and the fish fin (note for p. 92), it would be pedantic to inquire too closely into the detail of Miller's more elaborate metaphors, especially as this risks finding a misleading specificity in a phrase which was only loosely used for a specific rhetorical purpose. We have not found the expression 'line of existence' elsewhere in that form. However, Miller's friend Sir Thomas Dick Lauder wrote of the 'thermal line of existence' in a fascinating book on the Moray floods of 1829.[50] This was in the rather different sense of the altitudinal (but presumably also temperature-controlled) boundary seen in the distribution of snail species, analogous to the botanical 'treeline'.

P. 274: Cuvier ... happier world: Miller claims that Cuvier had speculated that humans would one day become extinct and that God would create new, higher beings to replace them. We cannot find any direct evidence of Cuvier having made such a speculation. However, an English translation and expansion of Cuvier's multi-volume zoological survey *Le Règne animal* ('The Animal Kingdom') does contain something very similar. In an introductory passage in his 'supplementary volume on the fossils', drawing on Cuvier's own *Ossemens fossiles* ('Fossil Bones'), the naturalist Edward Pidgeon had summarised the researches of Cuvier and his English followers, concluding:

> We find the simplest animals in the earliest secondary formations; as we ascend, the living structure grows more complicated – the organic development becomes more and more complete, until it terminates in man, the most perfect animal we behold. And shall we say that this march of creation has yet arrived at the farthest limit of its progress? Are the generative powers of nature exhausted, or can the Creator call no new beings from her fertile womb? We cannot say so. Revolution has succeeded revolution – races have been successively annihilated to give place to others. Other revolutions may yet succeed, and man, the self-styled lord of the creation, be swept from the surface of the earth, to give place to beings as much superior to him as he is to the most elevated of the brutes. The short experience of a few thousand years – a mere drop in the ocean of eternity – is insufficient to warrant a contrary conclusion. Still less will the contemplation of past creations, and the existing constitution of nature, justify the proud assumption that man is the sole end and object of the grand system of animal existence.[51]

It is not at all clear that this passage is a translation of anything in Cuvier's own work, and it goes against Cuvier's stated view in 1826 that humans represented the pinnacle of divine creativity – 'ce dernier et ce plus parfait ouvrage du Créateur' (this

---

50 Lauder 1830, pp. 235–36.
51 Pidgeon 1830, p. 39.

final and most perfect work of the Creator), as he put it in the third edition of his preliminary discourse to *Ossemens fossiles*.[52]

However, regardless of whether this passage was composed by Pidgeon or adapted from comments made by Cuvier in a forum as yet unidentified, its framing as a double question followed by an answer, and some details of its phrasing, suggest that Miller might have seen it and understood it to be a paraphrase of Cuvier's own views. This possibility is strengthened by numerous verbal echoes between Pidgeon's account and the last few pages of Miller's book: the 'march of nature' or 'march of creation' showing 'progression' or 'progress' towards the 'perfect' or 'perfection', the shared metaphor of 'lord of the creation', and the idea that the creative resources of God or Nature might be 'exhausted' or 'expended'.

We must also allow for the possibility that Miller heard about such views in conversation with Cuvier's sometime protégé Agassiz, perhaps attributed there to Cuvier. Finally, it should be noted that Miller's friend and Free Church ally John Fleming had extrapolated Cuvier's series of 'revolutions' into the future, hinting that another such cataclysm might bring about human extinction:

> if the same system of change continues to operate, (and it must do so while gravitation prevails,) the earth may become an unfit dwelling for the present tribes, and revolutions may take place, as extensive as those which living beings have already experienced.[53]

---

52 Cuvier 1826, p.172. We here cite this work in its separately published format, although it originated as the introductory part of *Ossemens fossiles*.

53 Fleming 1822, vol. II, p.104. On this passage, see Burns n.d., p.11; Burns 2007, p.211. Burns 2007 explores Fleming's complex attitude towards Cuvier's 'catastrophes', which was far from being as oppositional as is often assumed. As we discuss in pp.A157–64, Chambers appropriated Miller's own discussion of future superhumans in *Vestiges*. The verbal parallels there are with Miller rather than with Pidgeon, even though he drew a conclusion closer to Pidgeon's.

GLOSSARY

MILLER used the scientific terminology of 1841. Many of his terms are now obsolete and others have changed meaning, often in ways which do not allow simple translation into modern terminology. For instance, the identification of igneous rocks was based on gross properties of colour, texture, and so on which could be seen and felt in the hand, but modern geologists use different classifications based partly on microscopic study. We give broad equivalents for his scientific terms here, sufficient for most readers. We do not try to modernise biological classification in general, to explain in detail how any generic name as understood by Miller in the 1840s corresponds to the modern understanding of the same fossils, or to give the full formal taxonomic authorities for the names, for reasons explained elsewhere in this volume. We have not 'corrected' Miller's common habit of using roman script for genera (for which we use italic today). This is for practicality, and to respect the conventions of the time in their own context. We further outline differences with modern geological and palaeontological thought at the end of this book, including the equivalent chronologies and fish classifications.

acanthodians: a now-extinct group of fossil fishes, characterised among other things by the structure of their fins, each fin (except the tail fin) comprising a spine supporting a web of skin like the mainsail of a yacht

Acanthopterygii/acanthopterygians: the main grouping of living bony fishes, containing perch, mackerel and so on

aluminous: alum-containing (as of clays)

ammonite: extinct cephalopod mollusc with a coiled shell, similar to the modern pearly nautilus. Ammonites are common fossils in many Mesozoic rocks

anal: an anal fin is the (usually) single midline fin on the ventral midline of a fish, between the paired fins and the tail fin

annelid: a member of a group of worms which includes earthworms and ragworms

anthracite: a pure variety of coal

Arca: ark shell, a genus of bivalve mollusc

arenaceous: sandy, as in sandstone

argillaceous: clayey, clay-rich, as in mudstone and shale

Asaphus: a genus of trilobite

Ascidioida: sea squirts

Aspro: a genus of freshwater European fishes, including the streber and zingel

Avicula: a bivalve

Baculites: strictly, a genus of ammonite with a long shell in the shape of a curved cone rather than a spiral

basalt: a dark, hard igneous rock

belemnite: extinct cephalopod related to the modern squids and externally similar to them, but with a complex internal shell with air-filled buoyancy chambers. The belemnite shell had a solid bullet-shaped ending or 'guard'. Those guards were resistant to destruction after death and are common fossils

Bellerophon: a large sea snail with a bulbous shell

bituminous: rich in mineral pitch

bivalve: a member of the Bivalvia, a major group of molluscs, with paired left and right shells or 'valves'; examples are clams, oysters, and mussels. Very common fossils

bog-iron: a crude iron ore which forms in bogs today

brachiopods: marine shellfish superficially similar to bivalves but having top and bottom rather than left and right shells. Sometimes known as lamp-shells from the resemblance of some brachiopods to an ancient clay oil lamp. In the past, they were much more abundant and widely distributed than today

breccia: coarse rock made up of angular stones (which can include shells or even bones), cemented together naturally; the adjective is 'brecciated'

buckler: shield (military)

Burdiehouse Limestone: a limestone bed found in the Edinburgh region, of Lower Carboniferous age

calamite: a kind of giant horsetail, a common fossil in Coal Measures deposits

calcareo-argillaceous nodules: nodules of clayey limestone

calcareous: limy, rich in calcium carbonate

Cambrian: then the oldest known defined 'system' of sedimentary rocks; see Silurian

carbonic acid: carbon dioxide

Carboniferous: a grouping of strata including the main coal-bearing rocks; but 'carboniferous' could be a generic term meaning coal-bearing

Cardium: cockle

caudal: pertaining to the tail; a caudal fin is a tail fin, especially in fishes

centrum: the main, roughly cylindrical, body of a vertebra

cetaceae: cetaceans; whales and dolphins

Cephalaspis: a fossil jawless fish of the Old Red Sandstone, now extinct

cephalopods: a group of molluscs that includes octopuses and squids, and extinct forms such as belemnites, ammonites and orthoceratids

cervical: pertaining to the neck, e.g. a cervical vertebra

Cestracion: a genus of shark

chain of being: an old concept of order and plenitude within divine creation (pp. A67–68)

Chalk: the white limestone, often with embedded flints, famous from southern England and Yorkshire

chalybeate: of a spring, rich in dissolved iron minerals

chambered molluscs, chambered shells: shelled cephalopods (q.v.)

Cheiracanthus: a fossil acanthodian bony fish of the Old Red Sandstone, now extinct

Cheirolepis: a fossil bony fish of the Old Red Sandstone, now extinct

chelonian: one of the Chelonia, a group comprising tortoises, terrapins and turtles

chert: a rock composed of the mineral silica, but coarser and less homogeneous than flint

chevron bone: haemal arch, a bone in the tail of vertebrates; each haemal arch fits to the bottom of a vertebral centrum (q.v.), leaving a longitudinal aperture for a blood vessel running along the tail

Chimaera: a genus of holocephalan fishes, the group which includes the ratfishes or chimaeras

Coal Measures: a series of rocks, including coal, sandstone, fireclay and limestone, now attributed to the Carboniferous Period

Coccosteus: a fossil placoderm fish of the Old Red Sandstone, now extinct

condyle: the raised convex part of a skeletal joint

conglomerate: sedimentary rock composed of rounded, waterworn pebbles and boulders, naturally cemented together

copperas: metal sulphates, especially of iron

coprolite: fossil faecal dropping

Coral Rag: a series of strata within the Oolite (q.v.); today considered Middle Jurassic in age

coraline: coralline, (1) pertaining to coral, (2) a general term for the sort of hard aquatic algae and colonial animals which form a crust on rocks; on the sea shore

Cornstones: the middle part of the Old Red Sandstone system in England

cotemporary: contemporary

Crustacea: the group including lobsters, crabs, prawns, shrimps and relatives

ctenoid: relating to Louis Agassiz's classification of fishes based on the structure of their scales (see Appendix 4); ctenoid scales are thin and horny, one edge bearing a comblike fringe

Cucullaea: a bivalve mollusc

cuvy, rough-stemmed: cuvie or rough-stemmed tangle, a large seaweed similar to kelp

Cyclas: a freshwater bivalve

cycloid: relating to Louis Agassiz's classification of fishes based on the structure of their scales (see Appendix 4); cycloid scales are thin, horny, round and overlapping

den: wooded ravine (Scots) = northern English 'dene' (from Old English *denu*, valley)

diluvium: generic term (adjective **diluvial**) for certain superficial deposits such as soil, boulder clay, gravel and erratic stones, especially those today ascribed to the Ice Age glaciers and their melting (as was being debated in the 1840s). The word was a hangover from when those deposits were thought to be the sediment laid down by the 'Deluge' or Noah's Flood; in 1841, it was being replaced by 'drift'. See pp. A63–64

Diplacanthus: a fossil acanthodian fish of the Old Red Sandstone, now extinct

Diplopterus: a fossil bony fish of the Old Red Sandstone, now extinct; usually now named *Gyroptychius*

Dipterus: a fossil lungfish of the Old Red Sandstone, now extinct

dorsal: pertaining to the top or back, e.g. the dorsal fin or fins of a fish

dyke: molten rock may sometimes pass upwards through rock strata in more or less vertical fissures, where it may congeal to form, usually, basalt rock. If later exposed by erosion, the basalt is normally harder than the surrounding rock, and tends to be picked out as an upstanding wall-like feature, hence the name 'dyke' (Scots, field wall)

enamel: a highly mineralised hard tissue found in many vertebrates, usually in teeth, but also in scales and external bony plates in many fishes (either as true enamel or an analogous tissue). It often retains its high gloss in Old Red Sandstone fossil fishes

encrinites: the detached ossicles or individual elements from the skeletons of crinoids or sea lilies; can sometimes accumulate to form a major component of rock

exuviae: cast-off skins and shells; more generally, shells

factor: estate steward or manager (Scots)

fault: fracture and dislocation within a rock mass

feldspar: one of a group of crystalline minerals under this name; a major component of some igneous rocks such as granite (q.v.)

ferruginous: iron-rich

fire-clay: seat earth, the clayey layer often underlying a coal seam, and usually derived from the soil in which the plants forming the coal once grew

fluviatile: pertaining to a river, e.g. a sediment laid down by one

formation: here, a group of geological deposits referred to a common section of geological time; a subdivision of a system

frith: now 'firth', wide inlet of the sea (Scots)

fucus: *Fucus*, a genus of seaweed such as the familiar intertidal serrated wrack and bladderwrack. Miller appears to use it as a generic term for certain then poorly understood but superficially seaweed-like fossil plants of the Old Red Sandstone

galena: mineral, lead sulphide

ganoid, ganoidean: relating to Louis Agassiz's classification of fishes based on the structure of their scales (see Appendix 4); ganoid scales are large, bony, often enamelled and shiny, and often interlocking with each other

gastropods: a group of molluscs including the snails and slugs, on land and in water

Gault: a series of strata underlying the Chalk in England

German Ocean: North Sea

glottis: part of the windpipe to the lung(s)

Glyptolepis: a fossil lobe-finned bony fish of the Old Red Sandstone, now extinct

gneiss: a category of metamorphic rock, formed by the alteration of older rocks by heat and pressure when buried deep in the earth; with a more layered texture than granite but without the overtly platy, foliated texture of schist

granite: an igneous rock composed of visible crystals of quartz, feldspar and mica

graptolite: one of a group of extinct fossil animals common in 'transition' or 'grauwacke' rocks; of various shapes, such as those of feathers, tuning-forks or fans

grauwacke: also 'greywacke'. A lithological term for impure sandstone, but also often used to denote the sandstones and slates of a particular set of strata, equivalent to the modern Lower Palaeozoic

Great Oolite: a series of strata within the Oolite

Green-sand: strata overlying and underlying the Gault

gryphite: *Gryphaea* and similar fossil oysters, now popularly known as 'Devil's Toenails'

gunnel: butter-fish

haematic: probably an error for the iron ore haematite or the derived adjective haematitic

Helianthoida: the then name for sea anemones, corals and related animals

heterocercal: in tail fins, having a longer upper lobe than a lower one, as in sharks. Many bony fish such as cod have **homocercal** tails with equally sized upper and lower lobes symmetrical about the longitudinal axis of the body when seen from the side

hippurites: a coral-like animal (now considered a mollusc)

Holoptychius: a lobe-finned bony fish of the Old Red Sandstone, now extinct

horneblend: hornblende, a mineral which often occurs as coarse black crystals in igneous and metamorphic rock

hornstone: a fine-grained metamorphic rock

hyperstene: rock rich in the hard, dark mineral hypersthene

ichthyodorulite: fossil of a fish spine

ichthyolite: a fossil fish

ichthyology: the science of fishes

ichthyosaur: extinct marine reptile, like a (mammalian) dolphin in general appearance and presumed habits

inferior (of a rock formation): lower

intermaxillary bone: a bony element in the snout of some fishes

ironstone: rock rich in iron, iron ore

jasper: a siliceous mineral, typically coloured red and yellow

lacustrine: pertaining to a lake environment, such as sediments laid down in a lake

lateral line: the sensory line running along the flank of a fish

Lias (adjective **Liassic**): geological system, roughly equivalent to Lower Jurassic today

ligneous: woody

lignite: fossil wood, incompletely carbonised to true coal

limestone: a rock rich in lime, calcium carbonate

Lingula: a genus of brachiopod

Ludlow: a series of strata at the top of the Silurian

Lydianized schist: a hard form of schist

Magnesian Limestone: a series of strata overlying the Carboniferous

Malacopterygii: an old grouping of bony fishes, including salmon, pike and herring; not now current

mammifer: mammal

mammoth: an extinct fossil elephant, especially the woolly mammoth whose hair let it survive in polar climates during the cold periods of the Ice Ages

marl: clay mixed with limestone (can be hard or soft)

mastodon: another extinct elephant

Megalichthys: a fossil fish from the Carboniferous and uppermost Old Red Sandstone, in Miller's view; in fact conflating the fishes *Megalichthys* proper and *Rhizodus*

megatherium: giant ground-living sloth, now extinct

mica: a shining, often transparent, mineral, easily split into flat flakes

Modiola a bivalve similar to mussels

molluscs: the large zoological grouping of shellfishes such as clams, snails and squids

molluscous: pertaining to molluscs

Mountain Limestone: a series of Carboniferous strata below the Coal Measures

native: of metals, found naturally in a pure metallic state rather than having to be smelted from ore

Nereidina: a genus of ragworm, one of the annelid worms

New Red Sandstone (including **New Red Marls**): a series of strata overlying the Carboniferous

nodular limestone: limestone which occurs in nodules or concretions in a matrix such as clay shale

nummulites: the limy skeletons of certain large single-celled foraminiferan marine animals, often the size and shape of lentils or even small coins

occipital plates: bones of the rear portion of the head, here of fish

Onchus: an acanthodian fish, now extinct

Oolite (adjective **Oolitic**): at this time, a series of strata above the Lias; the name comes from the descriptive term 'oolitic' applied to the fish-roe-like appearance of some limestones from this tract, especially in the Cotswold Hills of England

operculum: hinged bony cover of the outflow of the gills in many fishes such as herring; **opercules**, small plates with the same function in some other fishes

organic remains: fossil animals and plants

Orthoceras, also *Orthoceratites*: a genus of cephalopod with a long straight conical shell, related to the modern pearly nautilus

oryctologist, oryctology: old terms for palaeontologist and palaeontology

osseous: bony

osseous fishes: bony fishes such as salmon, in contrast to cartilaginous fishes such as sharks

Osteolepis: a fossil bony fish of the Old Red Sandstone, now extinct

Ostrea: an oyster

Oxford Clay: a series of strata within the Oolite

pectoral: of or pertaining to the shoulder or shoulder joint; for instance, the pectoral fins of fishes

placoderm: a group of fossil fishes, now extinct, characterised by armour on the head and anterior body

placoid: relating to Louis Agassiz's classification of fishes based on the structure of their scales (see Appendix 4); placoid scales are plate-like, irregularly arranged on the skin and detached from each other, and often bear tubercles or rough points

Plutonic: of rocks, intruded into other rocks while still molten, and cooling and hardening there, at great depth in the earth, as opposed to volcanic rocks which also arise as molten rock but are erupted at the surface (in the sense of Lyell 1838, pp.14–16). More generally, the processes of the deep earth, such as those which injected granite and porphyry with the ensuing elevation of hills (as then understood)

Polynemus: paradise fish, a genus of freshwater fishes

porphyry: then, and today still informally, a term for an igneous rock containing distinct crystals scattered in a fine or glassy matrix

primary: by 1841, tending to mean the older basement rocks, both metamorphic (i.e. older rocks altered by heat and pressure deep down in the crust, such as gneiss) and igneous (i.e. formed by the intrusion of melted rock, such as granite), as opposed to secondary rocks (q.v.). Even then, 'primary' was an old-fashioned term, referring to those rocks which had been thought on older theories to be the most ancient, laid down directly from molten rock or a hot chemical-rich proto-sea

process: in an anatomical context, a protrusion on a bone

Pterichthyodes: modern genus name for *Pterichthys* (q.v.)

Pterichthys: a fossil placoderm fish of the Old Red Sandstone, now extinct

pyrites: the mineral pyrite, iron sulphide

quadrumana: literally, with four hands; an old zoological grouping comprising the monkeys and apes but excluding humans

quadrupeds: mammals, more specifically the 'four-footed' ones, thus excluding bats and primates (monkeys, apes and humans)

quartz: a pure form of silica; **quartzose**: rich in quartz

Quartz Rock: a stratum of quartz-rich rock occurring in parts of the Highlands

Radiata: the name of an old zoological grouping of (mainly) radially symmetrical invertebrates, including coelenterates (corals, sea anemones and jellyfishes) and echinoderms (such as starfishes and sea urchins)

Recent: the current period of geological time

rybat: dressed stone forming part of a window or doorway (Scots)

saliferous: rich in salt, containing salt (as in some sandstones)

sandstone: a rock composed of naturally cemented sand grains

sauroid fishes: fishes, such as the large predatory fishes found at Burdiehouse, then seen as resembling reptiles (particularly crocodiles)

schist (adjective **schistose/schistoze**): the modern meaning is of a rock composed of metamorphosed shale, and still showing the layering of the shale. In Miller's time, it also extended to other layered rocks generally, notably shales and the laminated muddy limestones of Caithness and Orkney

sea-pen: a colony of a pennatularian coral, which lives, not in a limy skeleton, but in a horny skeleton shaped somewhat like a feather quill pen, with the end (equivalent to the pen's nib) stuck in sea-floor mud. Miller was familiar with sea-pens netted by Cromarty fishermen (pp. 257–58)

secondary: old-fashioned term used by comparison with primary (q.v.), for rocks derived from other rocks, thus sedimentary rocks. Originally including older sedimentary rocks, as in Lyell 1830–33

shale: more or less hardened clayey rock which splits easily, especially horizontally along the bedding

Silurian: the system of strata underlying the Old Red Sandstone or Devonian. Its lower extent was a matter of major dispute in the 1830s and 1840s between Roderick Murchison and Adam Sedgwick, who had defined the Cambrian below (Rudwick 1985, Secord 1986)

siren: one of several North American species of salamander

slate: then, a general term for flat, layered and easily split rock, synonymous with schist in its older meaning; today a metamorphic form of clay shale altered by heat and pressure deep down in the earth

spar: generic term for crystalline mineral

sphærulites: fossil invertebrate, here probably a coral-like rudist mollusc

spiracle: in some fishes, the first gill slit which is reduced to a small hole. Miller uses it (unlike modern and at least some contemporary usage) to denote the exhalant branchial openings, arranged in a row on each side, from the gills in sharks and rays, as opposed to bony fishes such as salmon which have on each side a single crescentic opening covered by the bony flap of the operculum

stratified: arranged in layers or strata, esp. rocks, originally from bottom to top (though they may subsequently be tilted or folded)

sulphuret: sulphide; **sulphuret of hydrogen**, hydrogen sulphide gas, which smells of rotten eggs

superior (of a rock formation): upper, higher

suture: a junction between two bones, normally rigid

swine-stone: stone such as bituminous limestone which emits a foetid odour when struck by the hammer

system: here, and in geology of the time, a well-defined set of strata with characteristic fossils which could be recognised over a significant geographical area, and which represented a major division of geological time; for instance, the Silurian or the Old Red Sandstone

Terebratula: a genus of brachiopod or lamp-shell

tertiary: as coined by Lyell (1830–33), a major grouping of strata younger than and overlying the primary and secondary rocks (q.v.)

testacea: shelled molluscs such as snails, oysters, etc.

thorn-back: thornback ray (fish)

tilestones: slabby, splittable sandstones; **Tilestones** was the name for the lower part of the Old Red Sandstone in England

transition: originally rock strata intermediate between primary and secondary, in the older sense of those terms. By 1841, it had become a general term for (then still poorly understood) older sedimentary rocks below the Old Red Sandstone, which were often folded and somewhat metamorphosed, and often poor in fossils – all of which made it difficult to sort out their sequence

trap: here, usually basalt, a dark, hard, fine-grained rock, often, as here, cooled and solidified lava. Taken from the Swedish *trappa*, meaning a step on a stair (cognate with Scots 'trap'), via the characteristically staircase-like stepped terrain formed by such rocks, erosion having picked out the layers of successive lava flows. Such terrain is common in the Midland Valley and in the Inner Hebrides

travertin: travertine, a variety of limestone deposited from the water of springs which are rich in dissolved lime

trilobites: marine animals with segmented shells, superficially like large woodlice or slaters; now extinct

Trochus: top shell, a genus of marine snail

tuberculated: covered in small bumps or nodules

turbinated: in the shape of a decreasing spiral, as for instance in the shells of some snails or gastropods

Turbo: genus of large sea snail

Turrilites: a genus of ammonite drawn out into a conical spiral

Turritella: turret shell or tower shell, a genus of marine snail

ventral: of the belly, for instance of a fish; **ventral fins** (today often termed pelvic fins) were the rear of the two pairs of fins in many fishes

Venus: a clam-like bivalve

vertebral joint: here, a 'joint' in the sense of a single vertebra, rather than the articulation between two vertebrae

vitreous: glassy; **vitreous fracture**, breaking with a curved fracture surface, as glass does

Wealden: a series of strata above the Oolite

Wenlock: a series of strata within the Silurian as then understood

zoophite or **zoophyte**: a general term for animals such as corals, sponges, and so on, which live fixed like plants